Human Factors in the U.S. Railroad Industry

At the heart of the U.S. Railroad Industry are human operators who continue to make it work and operate. However, humans are susceptible to error, fatigue, and risky behaviors. Understanding how these impact the safety of the industry remains a crucial topic to be studied. *Human Factors in the U.S. Railroad Industry* investigates the human factor behind one of the world's biggest railway networks and synthesizes the body of research that has been produced by the Federal Railroad Administration (FRA) since 1993.

During that time, over 120 FRA reports were published on topics such as accident analysis, grade crossing safety and trespassing, working conditions and ergonomics, employee fatigue, safety culture, and track inspection. This book organizes this body of work into topic areas that correspond to Moray's Sociotechnical System (2006) to provide a comprehensive way to understand the relationship between policy, organizational culture, and system safety. It discusses tools such as risk exposure, signal detection theory, and program evaluation that have been applied to railroad projects and can be used to improve future railroad and transportation research. It provides a synthesis of the reports across topic areas such as fatigue, ergonomics, and safety culture that are currently important within the wider human factor transportation research community. The reader will develop an understanding of a "human-centered systems" approach to the railroad industry, which focuses on human capabilities and limitations with respect to human/system interfaces, operations, system integration, and organizational influences on safety.

Human Factors in the U.S. Railroad Industry will appeal to researchers, academics, students, and professionals interested in human factors, railway engineering, civil engineering, and industrial engineering.

Transportation Human Factors: Aerospace, Aviation, Maritime, Rail, and Road Series

Series Editor: Professor Neville A. Stanton, *University of Southampton, UK*

Systems Thinking in Practice: *Applications of the Event Analysis of Systemic Teamwork Method*
Paul Salmon, Neville A. Stanton, and Guy Walker

Individual Latent Error Detection (I-LED): *Making Systems Safer*
Justin R.E. Saward and Neville A. Stanton

Driver Distraction: *A Sociotechnical Systems Approach*
Kate J. Parnell, Neville A. Stanton, and Katherine L. Plant

Designing Interaction and Interfaces for Automated Vehicles: *User-Centred Ecological Design and Testing*
Neville Stanton, Kirsten M.A. Revell, and Patrick Langdon

Human-Automation Interaction Design: *Developing a Vehicle Automation Assistant*
Jediah R. Clark, Neville A. Stanton, and Kirsten Revell

Assisted Eco-Driving: *A Practical Guide to the Design and Testing of an Eco-Driving Assistance System (EDAS)*
Craig K. Allison, James M. Fleming, Xingd A. Yan, Roberto Lot, and Neville A. Stanton

Safety Matters: *An Introduction to Safety Science*
David O'Hare

Human Factors on the Flight Deck: *A Guide to Designing, Modelling and Evaluating*
Katie J. Parnell, Victoria A. Banks, Rachael A. Wynne, Neville A. Stanton, and Katherine L. Plant

Air Show Performers: *Safety, Risk Management and Psychological Factors*
Manolis Karachalios and Daniel Kwasi Adjekum

Driving Training for Automated Vehicles: *A Systems Approach*
Siobhán E. Merriman, Katherine L. Plant, Kirsten M. A. Revell, and Neville A. Stanton

Human Factors in the U.S. Railroad Industry
Thomas G. Raslear

For more information about this series, please visit: www.routledge.com/ Transportation-Human-Factors/book-series/CRCTRNHUMFACAER

Human Factors in the U.S. Railroad Industry

Thomas G. Raslear

CRC Press
Taylor & Francis Group
Boca Raton London New York

CRC Press is an imprint of the
Taylor & Francis Group, an **informa** business

Front cover image: Vineyard Perspective/Shutterstock

First edition published 2025
by CRC Press
2385 NW Executive Center Drive, Suite 320, Boca Raton FL 33431

and by CRC Press
4 Park Square, Milton Park, Abingdon, Oxon, OX14 4RN

CRC Press is an imprint of Taylor & Francis Group, LLC

© 2025 Thomas G. Raslear

Reasonable efforts have been made to publish reliable data and information, but the author and publisher cannot assume responsibility for the validity of all materials or the consequences of their use. The authors and publishers have attempted to trace the copyright holders of all material reproduced in this publication and apologize to copyright holders if permission to publish in this form has not been obtained. If any copyright material has not been acknowledged please write and let us know so we may rectify in any future reprint.

Except as permitted under U.S. Copyright Law, no part of this book may be reprinted, reproduced, transmitted, or utilized in any form by any electronic, mechanical, or other means, now known or hereafter invented, including photocopying, microfilming, and recording, or in any information storage or retrieval system, without written permission from the publishers.

For permission to photocopy or use material electronically from this work, access www.copyright. com or contact the Copyright Clearance Center, Inc. (CCC), 222 Rosewood Drive, Danvers, MA 01923, 978-750-8400. For works that are not available on CCC please contact mpkbookspermissions@ tandf.co.uk

Trademark notice: Product or corporate names may be trademarks or registered trademarks and are used only for identification and explanation without intent to infringe.

ISBN: 978-1-032-77140-3 (hbk)
ISBN: 978-1-032-81467-4 (pbk)
ISBN: 978-1-003-50001-8 (ebk)

DOI: 10.1201/9781003500018

Typeset in Times
by codeMantra

Contents

SECTION I   Introduction, Overview, and Context

SECTION II *Analytic Methods*

SECTION III *Grade Crossings and Trespassing*

SECTION IV Railroad Systems and Operations

Foreword

Grady C. Cothen*

When we think of the railroad industry, we often think of Amtrak and other intercity passenger services, present and future. As users of mass transportation provided on the general rail system of transportation, the Long Island Rail Road, Metra, Caltrain, or any of the 20 commuter railroads may come to mind. If we are oriented toward freight service, one of the six remaining Class I freight railroads may dominate our thinking. Our more local concerns may be pointed to one of the roughly 600 regional railroads, short line railroads, or historic and excursion railroads.

Behind all these organizations, of course, are suppliers, contractors, and fleet owners/managers and others who provide the locomotives and cars, specialized electronics and train control systems, advanced inspection technology, training packages, and a host of other products and services. These suppliers and contractors, overwhelmingly, have railroads as their primary or sole market. They constitute a part of the "railroad industry" as well.

What all these components have in common is a role, big or small, in making it possible for human beings to deliver rail service. All the parts must work together, presenting to the human actors in a way that permits them to achieve service requirements, the first of which is safety.

The public is a player too, navigating highway-rail grade crossings in automobiles, light trucks, heavy trucks, and all sorts of other vehicles—as well as by foot. This is an interface even more complex than any set of dials, levers, or displays or buttons on a locomotive cab console. Trespassers crossing live tracks take extreme risks that we would like to discourage, and those bent on suicide require a different set of countermeasures. But it is still all about human factors.

The railroad industry has benefited greatly from Human Factors research and guidance provided by experts at the Federal Railroad Administration (FRA), the Volpe National Transportation Center, various universities, and consulting organizations. The experts have drawn on research and practice in aviation, the military, and a variety of other domains internationally. In turn, ergonomists, operating officers, engineers, and others employed by the industry use the products of those efforts to make the railroad a safer and more sustainably productive enterprise.

To effect successful delivery of human factors learning, railroad managers and workers must be persuaded that change is both necessary and helpful. This may require regulatory muscle in some cases, but decision makers on the railroad can be nudged to adopt good ideas—particularly those that do not upend too many apple carts. The skill of applied research is being in the room and bringing participants to a more nuanced understanding of the problem and possible solutions.

As this book reveals, the awkward part of FRA-industry relationship is getting the foundation of knowledge together in a way that can be analyzed. Deriving the right data on which to find models and decisions can be a challenge. Of course, the researcher who is working for the agency with regulatory authority will be suspect to management, because management flexibility could be compromised. As a result, great stores of "proprietary" data are often unavailable for analysis in the public sphere. From the point of view of rail labor organizations, the collective bargaining status quo could be disrupted or lessons might be learned that disadvantage workers in litigation over accidents or injuries.

As an FRA lawyer and then senior executive at FRA (1973–2010), I got to witness first-hand the agency's Human Factors research and its practical application. When Dr. Thomas Raslear showed up in 1993, the agency kicked it up more than one notch. As readers of this volume will recognize, Tom is a research scientist with solid credentials. Arriving at FRA, he rapidly learned as much as he could about his new domain and commenced making contributions both to the literature and the practical business of railroad safety. He taught us about signal detection theory, cumulative fatigue, and the intersection between human performance and circadian rhythms. He showed us the impact of working conditions on safety outcomes, and he stressed the linkages between the effects of the environment on human health and job performance. He helped champion advances in safety culture, and he maintained an insistence on sound program evaluation—whatever the outcome might be.

Program evaluation matters, since not all good ideas can be made to work in practice, under all circumstances. For instance, as this is written, many are calling for the mandatory application of close call reporting programs on major freight railroads. The careful reader of this volume will learn why that would drive economic waste with little hope of good outcomes. By contrast, where workers and managers want to make the program work, Tom and his colleagues showed that it can be adapted to their needs with outstanding results.

During his stint at FRA, Tom worked with the team, which included researchers and regulators, to show how theory could be validated with data from a variety of sources. He skewered the pretenders with what I am advised (not being qualified to judge) is a tremendous understanding of applied statistics. He made colleagues into friends and collaborators. His sense of humor and patent (although not always patient) goodwill endeared him to railroad employees and their representatives.

Remarkably, Tom joined the roles of supervisor, modally connected partner within USDOT, program manager, and hands-on researcher in his personal portfolio. This had been done previously, but not often with the energy and constructive aggressiveness that Tom brought to the tasks at hand. This was not expected, and it was not required to achieve good performance reviews. It came with the person, not the job. His colleagues showed their readiness to embrace similar responsibilities.

This book is the story of Tom's human factors journey through railroad safety, gathering lessons from his extensive publications and those of his colleagues and providing context to explain what the agency was able to accomplish and what shortcomings sometimes held it back. It has sufficient analytic heft to satisfy academic readers, and then some. But it is also possible to leap the formulas and go for the clearly stated descriptions of objectives, methods, and findings. It is not possible to read this book without learning something new, even for those of us who lived through a good part of it.

Like risk reduction elsewhere, achieving railroad safety is a process of clearly identifying and chipping away at one knotty problem after another. To do it justice, you must want to do it. Tom Raslear has shown us through this book that, while he may be retired from government service, he is not through teaching us about human factors—and not through trying to make us safer.

NOTE

* Grady C. Cothen Jr. retired in 2010 from the Federal Railroad Administration after 36 years in various positions: Deputy Associate Administrator for Safety Standards and Program Development (1994–2010); Associate Administrator for Safety (1991–1994); and Special Assistant to the Chief Counsel (1980–1991). Prior to that, he served as a trial attorney in the Enforcement Division of the FRA Office of Chief Counsel. Cothen was also Acting Associate Administrator for Policy from 1986 to 1988. Following his retirement from Federal service, he provided consulting services to Amtrak and a major rail transit system through 2016. Now fully retired, he has continued to write on behalf of railroad safety and present before various expert bodies, including the rail subcommittee of the House Committee on Transportation and Infrastructure.

Preface

This book is intended to be an overview of Human Factors research conducted by and for the Federal Railroad Administration. The time frame is mostly focused on my tenure as an engineering psychologist at FRA (1993–2015), but also includes background information about research conducted prior to and following that time. It is designed to provide information about critical human factors-related safety issues to railroad management, railroad union officials and members, and FRA staff. All of the topics covered in this book have relevance to other industries and modes of transportation, so the book can serve as a resource for other human factors researchers and students. The original reports are available for free on the FRA's website (https://railroads.dot.gov/elibrary-search) but are not widely referenced. Research performed and reported by federal agencies is rarely cited in the peer-reviewed psychology/human factors literature even though the research topics, methods, and results are relevant and important to that literature. For instance, a recent report by Hamedani, Markus, Hetey, and Eberhardt (2024)[*] in the American Psychological Association's flagship journal, *American Psychologist*, discusses intentional culture change in organizations. They say that "Calls for culture change abound. Headlines regularly feature calls to change the 'broken' or 'toxic' cultures of the police, the workplace, U.S. politics, and more, and norms and practices across society are hotly debated. The proposed fix is 'culture change.' But what is culture change? How does it work? And can it be effective?" (p. 384). Their report cites over 180 references, but no government reports. The FRA, over a period of more than 15 years, conducted and published research on culture change in the railroad industry (see Chapter 9). Three different methods were used (safety rules revision, behavior-based safety, and confidential close call reporting), and the outcomes were rigorously evaluated to document their effectiveness. This book attempts to bring needed attention to research by federal agencies.

The book is organized into 10 chapters in four sections. Section I, "Introduction, Overview, and Context," contains Chapter 1, an introduction and overview of Human Factors research at FRA including the relationship of the program structure to the sociotechnical system; and Chapter 2, the social, legal, and policy context of FRA research. Section II, "Analytic Methods," contains Chapter 3, a discussion of data issues concerning the FRA accident/incident database, accident frequency and cost, and risk; Chapter 4, a discussion of exposure in the analysis of accident frequency data. Exposure in this context refers to the extent to which employees are subjected to conditions which may affect risk; and Chapter 5, an overview of signal detection theory which is used to analyze motorist behavior at grade crossings, track inspection, train conspicuity, train horn audibility, and wayside track signals. Section III, "Grade Crossings and Trespassing," contains Chapter 6, which describes a signal detection theory model of motorist behavior at the grade crossing and efforts to understand and prevent trespassing. Section IV, "Railroad Systems and Operations," contains Chapter 7, a discussion of railroad employee fatigue and hours of service regulations. A biomathematical model of fatigue based on time on duty and circadian rhythms is described; Chapter 8, a discussion of working conditions and ergonomics of railroad work with a primary focus on the physical system of the locomotive cab, the environment in which it operates, and the effect of that physical system on the behavior of individual workers and the teams that they form; Chapter 9, a discussion of three safety culture programs: safety rules revision, behavior-based safety, and close call reporting systems. The effectiveness of the programs and conditions to implement them are described; and Chapter 10, a discussion of track inspection by humans and automation. A theory of the ideal observer is developed to determine the speed at which humans can detect track defects, and signal detection theory is used to compare human and automated track inspection.

NOTE

* Hamedani, M. G., Markus, H. R., Hetey, R. C., and Eberhardt, J. L. (2024). We built this culture (so we can change it): Seven principles for intentional culture change. *American Psychologist*, 79, 384–402.

Acknowledgments

The research I have described here is the product of the efforts, direct and indirect, of many individuals in many organizations. At the Federal Railroad Administration (FRA), much of the research planning and structure of the Human Factors program was conceived with the invaluable assistance of Michael Coplen. Program evaluation of many of the research projects was entirely a result of Mike's dogged insistence that projects needed to document their degree of effectiveness. Claire Orth, as my first immediate supervisor at FRA, allowed me the freedom and support to pursue projects that did not exactly fit the prescribed operating plan. Garold Thomas mentored me as a newcomer to the railroad industry and the FRA. Many of his aphorisms guided me in my career at FRA as a researcher who was expected to be responsive to the Office of Safety and the Administrator, particularly "If they want it bad, that's how they'll get it." Scott Kaye provided major support for the research program on hours of service and employee fatigue. Grady Cothen and Jo Strang supported research programs that provided data to support the development of regulations on locomotive cab working conditions, hours of service, and safety culture. Brenda Hattery, Christine Beyers, Jeff Horn, and Colleen Brenan served with me on several Railroad Safety Advisory Committees that used Human Factors research to formulate regulations. They helped me understand the technical, economic, and legal underpinnings of regulations that research needed to support. Dennis Yachechak provided insights and details concerning carrier safety rules in several research projects, and John Conklin provided assistance with access to safety data on projects concerned with the remote control of yard locomotives.

At the Volpe Center, many research projects were directly supported by Don Sussman and Jordan Multer. Their knowledge of the railroad industry and human factors was essential to the excellence of many research projects that FRA funded through Volpe. Jordan, in particular, was key to the development of the Confidential Close Call Reporting System project. Joyce Ranney, J. K. Pollard, Eric Nadler, and Catherine McInnis provided exceptional research performances in the execution of many research projects. Michelle Yeh provided essential expertise to the further development of signal detection theory models for motorist grade crossing behavior.

Several contractors were engaged by Volpe to assist in the conduct of research projects. They include Johnathan Morell, Dennis Bley, John Wreathall, Tom Sheridan, Julie Hile, and Emilie Roth. They provided expertise in areas that supplemented Volpe personnel. Many projects would not have been possible without them.

FRA directly funded contractors with human factors expertise. They include Syntek Corporation (formerly Fulcrum) and Foster Miller Corporation. Vijay Kohli of Syntek provided research support for a variety of projects to introduce new technology into railroad operations. Judith Gertler of Foster Miller conducted most of the research on railroad employee work-rest diaries which is partly the basis of the latest hours of service regulations in the industry. Steven Hursh of the Institutes for Behavior Resources (IBR) formulated, validated, and calibrated a fatigue model using railroad accident data. That model is widely used (ironically) in the aviation industry today.

The views expressed herein are those of the author and do not necessarily reflect the views of the IBR or of any of the organizations or individuals named above. Any errors are the sole responsibility of the author.

About the Author

Thomas G. Raslear is an Independent Human Factors Consultant based in Maryland, USA. He has over 40 years of experience in experimental psychology and Human Factors research. For 23 years, he was based at the Federal Railroad Administration (FRA) until his retirement. At FRA, he was Chief of the Human Factors Research Division in the Office of Research and Development. He is an internationally known expert in fatigue in the railroad industry and has published extensively on the effects of work schedules on fatigue and accidents. Tom was a member and former chair of the U.S. Department of Transportation Human Factors Coordinating Committee and a member of the Transportation Research Board Standing Committee on Railroad Operational Safety (AR070). He is also an internationally known expert on grade crossing human factors and accident causation and has published extensively on the use of Signal Detection Theory in the analysis of motorist behavior at grade crossings.

Section I

Introduction, Overview, and Context

1 Introduction and Overview

The purpose of this book is to synthesize the body of research that has been produced by the Federal Railroad Administration (FRA) on Human Factors in the railroad industry between 1993 and 2015. During that period of time, over 120 FRA reports were published on topics that range from accident analysis to task analysis. The research was funded by Congress and managed by the Human Factors Research Division and its predecessor, the Equipment and Operating Practices Division. The research was performed by contractors and by government researchers at the John A. Volpe National Transportation Systems Center (Volpe) and FRA. The research is easily catalogued into topics that most Human Factors researchers would recognize, but the reports stand as individual works of interest to the railroad industry that are often not related in a systematic way to the larger Human Factors literature. Moreover, these government reports are rarely cited in the Human Factors literature, perhaps because they are not as accessible as journal articles and academic books, or because there is a prejudice in academic circles against science that is not "peer-reviewed." This book organizes this body of work into topic areas that correspond to Moray's Sociotechnical System (2006). This provides an opportunity to display the breadth and depth of the research that FRA produced. It also allows a synthesis of the reports across topic areas that are currently important within the Human Factors research community.

HUMAN FACTORS

This book is organized as Human Factors topics pertinent to railroad safety nested within the sociotechnical system. This follows the programmatic organization of FRA's Human Factors Division's research program from 1993 to 2015.

The term "Human Factors" refers to the design of systems, subsystems, and components for human use. This entails the use of "human-centered systems" design principles that respect human capabilities, limitations, characteristics, behavior, and motivation and recognizes that systems, subsystems, and components can only be as effective as the humans who must use them. The goal of human factors requirements and concepts in system design is to increase the effectiveness and efficiency of work; the reduction of human error, fatigue, and stress; and enhance safety. Since the implementation of any new system, subsystem, or component can directly or indirectly change the nature of tasks that humans perform, both negative and positive consequences of implementation should be considered in design. For this reason, Human Factors research is a cross-cutting issue in many other areas of research and technology development. In the railroad industry, as new technologies, such as digital communications, positive train control, automation, and electronically controlled

DOI: 10.1201/9781003500018-2

pneumatic brakes, are developed it is essential that the human factors consequences of these technologies are carefully considered.

FRA's Human Factors research program can be seen as part of a "human-centered systems" approach, which focuses on human capabilities and limitations with respect to human/system interfaces, operations, system integration, and organizational influences on safety. The goal of an increased attention to human performance and behavior using a systems approach is to reduce crashes, loss of life, injuries, property damage, and resultant personal and financial costs.

SOCIOTECHNICAL SYSTEMS APPROACH

The Human Factors Program follows a systems model based on Neville Moray's (1994, 2006) structure of sociotechnical systems, shown in Figure 1.1. In this model of nested influences, each layer encompasses the content of inner layers. Outer layers of the framework contain elements of inner layers and influence the inner layers, giving the program elements a great deal of interconnectivity. Leveson (2011) has extended this model to the analysis of systems safety which places the FRA Human Factors research in an appropriate context for systems safety management.

Beginning with the inner layers, the core elements of the sociotechnical systems model include: (1) the physical systems, (2) individual/team behavior, (3) organizational/management infrastructure, and (4) social, legal, and policy context. Individuals rely on the physical system (e.g., a locomotive) to perform their jobs. The design of physical system characteristics (e.g., displays and controls) affects how individuals interact with the system to perform their jobs. Changes in physical systems cause changes in how jobs are performed and directly affect safety. Individuals perform their jobs within the context of personal (biological and psychological), environmental, and social conditions that also affect job performance and safety. So too, teams of individuals communicate, coordinate, and cooperate to perform inter-related tasks on various physical systems to achieve a common goal (e.g., move a train between two locations). This teamwork is performed within the context of group dynamics

Elements of a Socio-Technical System

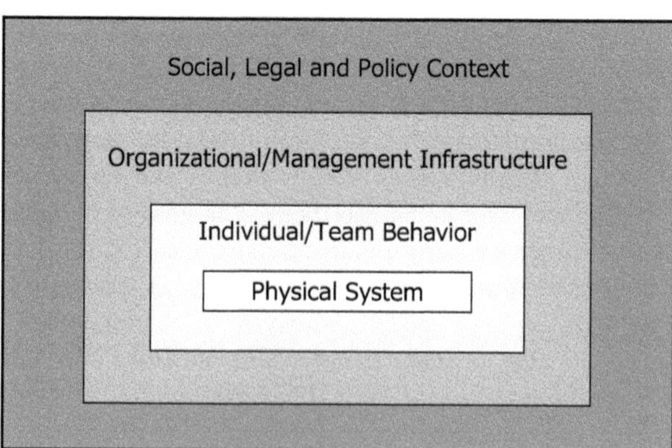

FIGURE 1.1   Elements of a sociotechnical system.

that affect task performance, safety, and goal attainment. Teams set goals, engage in communication, coordination, and cooperation to meet those goals in accordance with organizational values and assumptions about appropriate goals (e.g., productivity vs. safety), communication, coordination, and cooperation. Leadership in the organizational/management infrastructure lays the foundation for organizational safety climate and culture, which is represented by the underlying safety norms and common practices within the organization, often described simply as "the way things are done around here." This resulting organizational culture and safety climate affect team and individual job performance, physical system design, and safety. Finally, the social, legal, and policy context within and outside the industry influence organizational leadership and decision-making, creating the outermost layer of a sociotechnical system from which organizational safety culture and climate ultimately plays out.

FRA HUMAN FACTORS R&D PROGRAM STRUCTURE

As shown in Figure 1.2, FRA's Human Factors R&D Program was structured into two broad program areas, each with subordinate subprogram areas of research. The broad program areas include: Railroad Systems & Operations, and Grade Crossings & Trespassers. Under each of these broad program areas are three subprograms:

Technology, including Automation and Systems Design;
Railway Worker/Operator Performance, Safety, and Health; and
Organizational Culture and Safety Performance.

The overall R&D program also includes two supporting program areas under Strategic Planning:

Technology Transfer
Program Evaluation.

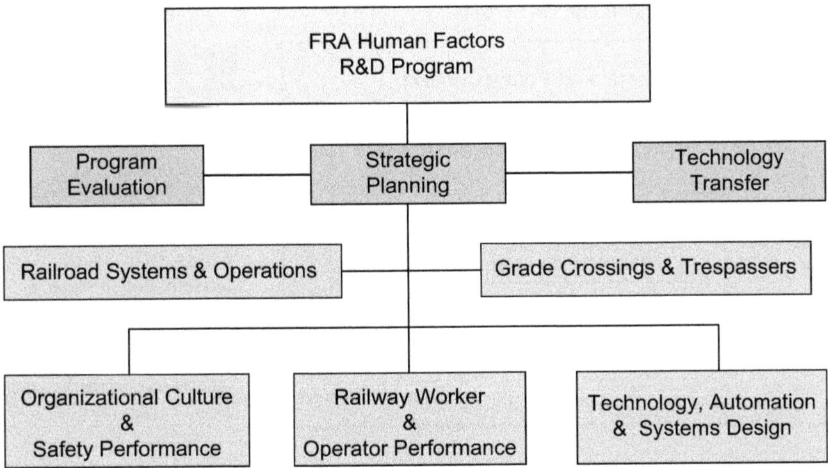

FIGURE 1.2 Organizational structure of the FRA Human Factors research program.

Each broad program area and subprogram area is supported by specific projects. Supporting program areas help knit the entire program together. Ongoing strategic planning includes the prioritization and selection of research projects, project coordination, and development of the program activities. Periodic context assessments of the social, legal, and policy barriers in the industry are also conducted, which provide the necessary contextual understanding needed for building stakeholder commitment, collaborative research and demonstration partnerships, and long-term sustainability in programmatic Human Factors research. Program evaluation and technology transfer initiatives help increase the feasibility, utilization, impact, and effectiveness of the program.

FRA's HUMAN FACTORS R&D PROGRAM IN A SOCIOTECHNICAL SYSTEM

FRA's Human Factors R&D Program has been structured to correspond to the layers identified in the sociotechnical system (see Figure 1.3). This balanced, multilayered approach to enhancing the safety of railroad operations includes programmatic areas of research in all layers of the sociotechnical system to maximize safety impact at the broadest levels across the industry.

OBJECTIVES

Evaluation, Technology Transfer, and Strategic Planning

This subprogram element focuses on the selection of projects, evaluation of program utility and effectiveness, and product dissemination. Because human factors-related accidents and injuries account for such a large proportion of overall incidents, it is imperative that periodic evaluations be conducted to assess program strengths and

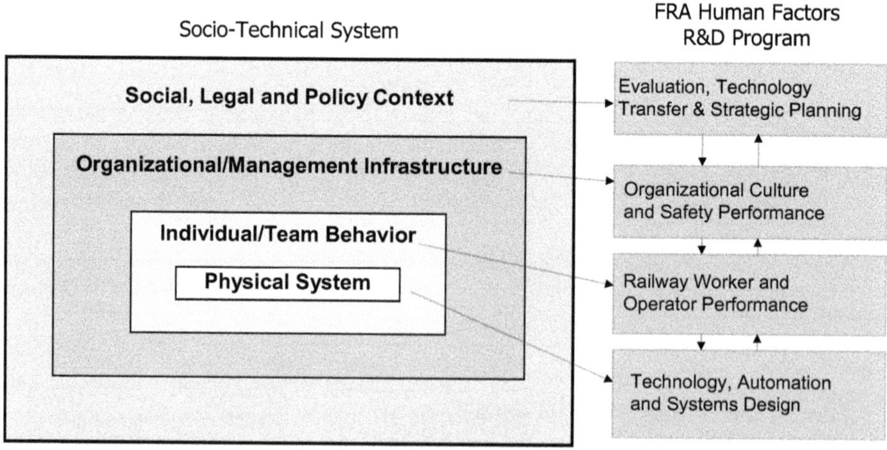

FIGURE 1.3 Correspondence between the FRA Human Factors research program and the sociotechnical system.

weaknesses and provide direction for future improvement. Both internal and external factors that affect or influence the overall success of the Human Factors Program are included in that assessment. This subprogram assesses the need for Human Factors research in various railroad operations; develops specific performance goals and objectives based on the overall needs of the industry; develops a plan for implementing recommended improvements that will help achieve these program goals and objectives; develops performance indicators to be used in assessing the outcomes of the Human Factors Program; and improves the overall effectiveness of the Human Factors Program.

RAILROAD SYSTEMS & OPERATIONS

Organizational Culture and Safety Performance

Organizational culture, defined as shared values, norms, and perceptions that are expressed as common expectations, assumptions, and views of rationality within an organization, plays a critical role in safety. Organizations with a positive safety culture are characterized by communications founded on mutual trust, by shared perceptions of the importance of safety, and by confidence in the efficacy of preventive measures. This subprogram focuses on programs to enhance railroad safety by encouraging the development of a positive safety culture within the railroad industry by addressing issues such as what gets reported when errors occur; how is blame apportioned; and what lessons are learned from errors.

Railway Worker & Operator Performance

Individuals and groups (teams) of workers perform safety-critical jobs in the railroad industry under a variety of personal (age, sleep deprivation, motivation, memory, etc.), environmental (noise, temperature, vibration, etc.), and social (status, role, etc.) conditions that may affect job performance and safety. This subprogram examines these factors to identify those that have significant impacts on job performance and safety and to suggest strategies to enhance safety and job performance.

Technology, Automation, & Systems Design

The introduction of new communications and computer technology in the railroad industry will change how workers in railroad operations perform their jobs. Automation and the management and control of information are becoming essential components of railroad operations. This subprogram examines the safety implications of new technology and automation from a human-centered design perspective. In general, systems design issues are also addressed in this subprogram, regardless of the recency of the technology involved. The primary goal of the subprogram is to ensure that safety is enhanced, not degraded, by new technology and automation. Prototypes of new systems are also designed and tested in order to benchmark the unintended human factors consequences of new technology and to provide viable solutions for their resolution.

GRADE CROSSINGS & TRESPASSERS

As noted above, the Grade Crossings & Trespassers program area has the same subprogram elements as the Railroad Systems and Operations program area. However, because the physical system is more complex (rail and highway infrastructure and vehicles), projects in this program area have focused on the physical system and the interaction of motorists with that physical system. This program area also focuses on why motorists and commercial vehicle operators would take risks such as driving around or through lowered gates (motivation, expectations, and perceptions) and on changes to grade crossing and train systems design to enhance safety.

Trespassers now account for the highest number of fatalities in the railroad industry. In many industrialized countries, suicides account for a large number of trespasser fatalities, and the proximity of mental health facilities to rail infrastructure accounts for many "hot spots" in trespasser incidents. However, there are many reasons for trespassing on railroad rights-of-way, some of which involve recreation, community access to services, and media portrayals of such trespassing as a cultural rite of passage.

Organizational Culture & Safety Performance

In the case of grade crossings, the regional and national culture and institutions play a critical role in what is considered to be acceptable and safe motorist behavior at crossings. State and local governments send a message about safety to motorists when crossing violations are tolerated and enforcement is inappropriate to the risk involved. Coordination between state agencies and local governments regarding traffic signals, road markings, and signage is essential to properly communicate risk to the motorist public. Courts that disdain the use of technology in the enhancement of enforcement (e.g., photo enforcement) also fail to communicate the reality of the risk at grade crossings. Whereas in many European countries, grade crossing accidents are taken as direct evidence in court of a lack of care taken by the motorist, in the United States the opposite is usually true. This is a difficult area in which to effect change and is probably the biggest challenge.

Similarly, national and regional culture and institutions play a critical role in what is considered to be acceptable and safe behavior around railroad property. Trespassing on railroad rights-of-way may be encouraged by portrayals in the media and in advertisements of such behavior as "cool" and as a normal part of growing up. The acceptability of recreating on rail property (e.g., diving off railroad trestles) clearly has a cultural component which should be changed to discourage trespassing.

Railway Worker & Operator Performance[1]

Motorists perform safety-critical tasks in driving their vehicles under a variety of personal (age, sleep deprivation, motivation, memory, etc.), environmental (noise, temperature, vibration, etc.), and social (status, role, etc.) conditions that may affect their behavior at the grade crossing and safety. This subprogram examines these factors to identify those that have significant impacts on grade crossing behavior and safety and to suggest strategies to enhance safety and performance.

Similarly, trespassers engage in activities on rail property for a variety of reasons and under specific circumstances. Knowledge of trespasser characteristics, motivations, and circumstances can contribute to the development of mitigation strategies and the prevention of trespassing.

Technology, Automation, & Systems Design

The introduction of new communications and computer technology will change how motorists interact with grade crossings and trespassers interact with railroad rights-of-way. Automation and the management and control of information are becoming essential components of motorist behavior and railroad operations. This subprogram examines the safety implications of new technology and automation from a human-centered design perspective. In general, systems design issues are also addressed in this subprogram, regardless of the recency of the technology involved. The primary goal of the subprogram is to ensure that safety is enhanced, not degraded, by new technology and automation. Prototypes of new systems are also designed and tested in order to benchmark the unintended human factors consequences of new technology and to provide viable solutions for their resolution.

ROLE OF THE HUMAN FACTORS RESEARCH PROGRAM WITHIN THE FRA

WHY IS HUMAN FACTORS R&D IMPORTANT?

Since 1985, human factors accidents have accounted for approximately one-third of all railroad accidents and half of all yard accidents. In 2014, 692 human factor-caused train accidents occurred, which was 38.5% of the total accidents. The number of train accidents from 1975 to 2014 by cause is shown in Figure 1.4. Figure 1.5 shows the proportion of train accidents over the same time period. It is clear that human factors are a major contributor to train accidents and need to be seriously addressed.

The reduction of human factors accidents requires examination of current railroad operating practices, industry trends, and anticipation of the future safety of the industry as practices evolve. Yard and terminal accidents may be caused by shortcomings in operating practices that include the methods and materials that are used to train and test employees in the performance of their jobs, the methods and materials that are used to perform specific jobs and tasks, the rules that govern job and task performance, and the general interaction of employees with the job environment and supervisors. Operating practices can result in human factors accidents for a variety of reasons. For instance, lack of training may cause accidents because the training methods are inadequate or inappropriate, because the training materials lack readability or are inappropriate for the education level of the employees, or because the testing methods are lax. Disproportionate numbers of human factors accidents in specific job categories or environments currently provide the best indication that operating practices should be critically examined (Figure 1.5).

Trespassers now account for the highest number of fatalities in the railroad industry. In 2014 there were 490 trespassing fatalities and 415 injuries. Grade crossings present a major hazard to motor vehicle drivers, as well as pedestrians, and are

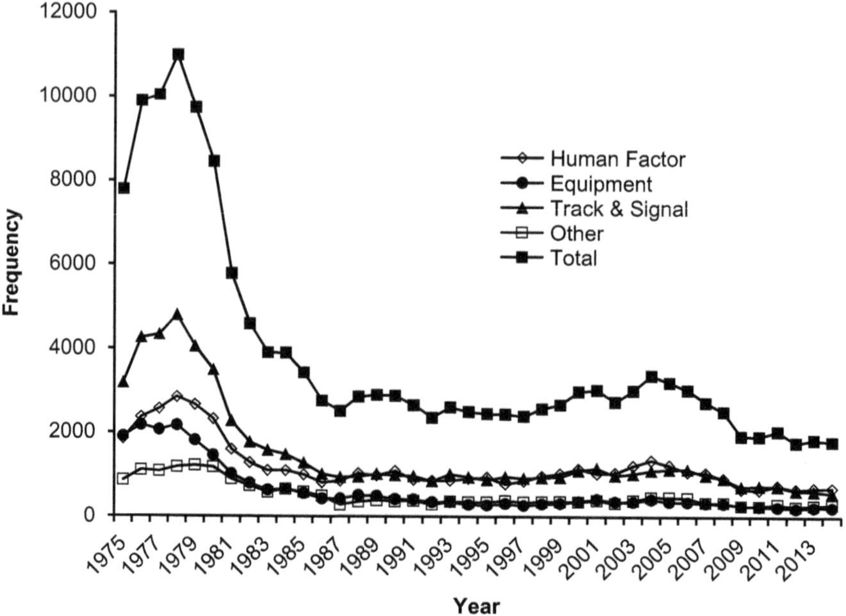

FIGURE 1.4   Train accident frequency from 1975 to 2014 by cause.

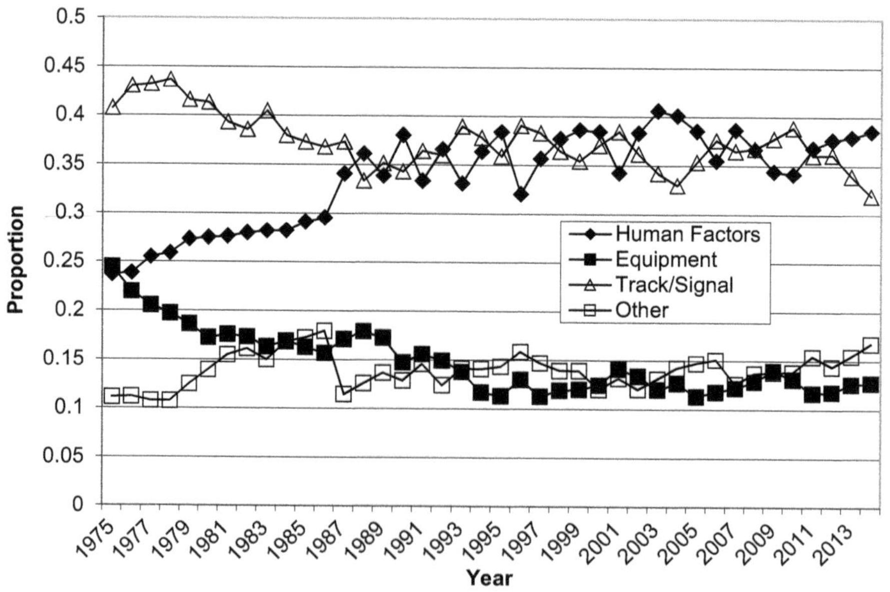

FIGURE 1.5   Proportion of train accidents from 1975 to 2014 by cause.

the second highest cause of fatalities and injuries in the railroad industry. In 2014, there were a total of 1,966 incidents at public crossings, resulting in 240 fatalities and 729 injuries. Many grade crossing accidents are directly due to motorist and commercial vehicle operator behavior. In 2007 (January to November), the majority of accidents occurred at grade crossings with active warning devices (e.g., gates and flashing lights). Historically, 21% of such accidents were caused by motorists and commercial vehicle operators not stopping, and, in many situations the flashing red lights were ignored. Historically, in 10% of accidents, the motorists and commercial vehicle operators actually went around or through lowered gates. Why motorists and commercial vehicle operators would take such risks is unknown and was examined through several research projects over several years.

RELATIONSHIP TO OTHER FRA OFFICES

The Human Factors Research Division is part of the Office of Research and Development in the Office of Railroad Policy and Development. The Human Factors Research Division provides direct support to FRA's Office of Safety, Chief Counsel, and other research divisions in the Office of Research and Development. The Operating Practices Division in the Office of Safety is a direct parallel to the Human Factors Research Division in the Office of Research and Development. It has long been recognized that human factors play a significant role in accident causation. Of the 389 accident cause codes used by FRA in its accident database, 30% (118) are human factors cause codes (Sussman and Raslear, 2008). Support to the Office of Safety and Chief Counsel is often in the form of research to provide a scientific basis for new regulations at the request of the Office of Safety and/or Chief Counsel. Consensus rulemaking directly involves Human Factors Research Division staff in technical meetings of the Railroad Safety Advisory Committee (RSAC, see Cothen et al., 2005; Federal Railroad Administration, 2001, for a description of the process). Recent changes in regulations on hours of service provide a good example of how Human Factors research has provided a scientific basis for FRA rules (see Raslear and Brennan, 2013 and Chapter 7). Traditional rulemaking, sometimes called "hear and decide" (Cothen et al., 2005), also can involve research staff to assist in gathering human factors information pertinent to the topic under consideration.

Since the Human Factors Research Division monitors safety trends and the development of technology in the railroad industry, the division also performs research that attempts to anticipate the knowledge needs of the FRA. In some cases, that involves the development of technology to assist staff in the Office of Safety in the performance of their jobs, or in the development and testing of programs to enhance safety in the industry (see Discussion of Safety Culture, Chapter 9).

Apart from direct research support, the Human Factors Research Division also provides technical advice to other FRA offices and research divisions regarding human factors, research design, statistical analysis, and probability. During the time period covered by this book, the Human Factors Research Division had only four staff (chief and three program managers), so many technical issues were referred to experts at Volpe. A complete, official description of the functions of Human Factors Research Division within the FRA can be found in the Federal Railroad Administration Organization Manual (Federal Railroad Administration, July 11, 2011).

RELATIONSHIP TO OTHER DEPARTMENT OF TRANSPORTATION AGENCIES

Many of the human factors issues that concern FRA are common to other transportation modes. The Human Factors Research Division participates in the Department of Transportation's (DOT) Human Factors Coordinating Committee (HFCC) which meets regularly to share information about research projects and to advocate for the application of human factors science in transportation (see http://hfcc.dot.gov for Vision, Mission, and Goals statements). In addition to DOT agencies, HFCC now includes staff from the National Transportation Safety Board, Centers for Disease Control and Prevention, the U.S. Coast Guard, U.S. Airforce, U.S. Army, and U.S. Navy.

RELATIONSHIP WITH THE RAILROAD INDUSTRY

The railroad industry is not a single entity. Consequently, the relationship between the FRA and the industry it regulates is multifaceted and complex. The major constituents of the industry are the carriers, railroad employees, and suppliers.

The carriers consist of Class I, Class II, and Class III railroads. The Surface Transportation Board (STB) has designated railroad class on the basis of a railroad's operating revenue. STB is a federal agency responsible for the economic regulation of the railroad industry. According to Wikipedia, a Class I railroad has operating revenue of more than $504,863,244, a Class II railroad has operating revenue of $40,384,263, and a Class III railroad has operating revenue of less than $40,384,263. These were the thresholds set in 2019. Operating revenue thresholds are adjusted each year for inflation.

The Class I railroads as of 2011 include BNSF Railway, Canadian Pacific Railway, Canadian National Railway, CSX Transportation, Kansas City Southern Railway, Norfolk Southern Railway, Union Pacific Railroad, and Amtrak. Class I railroads haul commodities over long distances. The Class I railroads have the majority of trackage (approximately 70%) and railroad employees (approximately 89%) according to the Association of American Railroads (AAR, the trade association for Class I railroads). For this reason, most of the research performed by FRA has focused on Class I railroads.

Class II and III railroads are the regional and short line railroads. The American Short Line and Regional Railroads Association (ASLRRA) is the trade association for 545 Class II and Class III railroads. The regional railroads are line haul railroads that have at least 350 miles of track (AAR, 2005). The remaining railroads are short line and terminal railroads. Included in these two classes are passenger railroads and tourist railroads. The passenger railroads include railroads that are operated by states for commuters such as the Long Island Railroad and New Jersey Transit. The American Public Transportation Association (APTA) is the trade association for commuter and passenger railroads. While APTA members also include subways and light rail, transit systems such as Washington, DCs Metro and New York City's subway system which are not connected to the U.S. railroad network are not subject to FRA regulation and have not been the subject of FRA research.

Railroad employees work in a variety of jobs. According to the AAR (2021), railroad employees include executives, officials, and staff assistants (6.3%); professionals and administrative staff (8.8%); maintenance of way and structures (24.7%); maintenance of equipment and stores (16.9%); transportation, train and engine (39.2%); and transportation, other than train and engine (4.3%). FRA Human Factors research has been concerned almost exclusively with those employees who operate and maintain the trains, equipment, infrastructure, and tracks. The majority of these employees are unionized. Consequently, railroad labor unions are important stakeholders in FRA Human Factors research. There are many labor unions that represent railroad employees, but not all have been active in FRA regulatory affairs or research activities. Labor unions that have interacted with the FRA Human Factors research program include the Brotherhood of Locomotive Engineers and Trainmen (BLET), the United Transportation Union (UTU, now incorporated into the International Association of Sheet Metal, Air, Rail and Transportation Workers, also known as SMART), the Brotherhood of Railroad Signalmen (BRS), the Brotherhood of Maintenance of Way Employees (BMWE), the American Train Dispatchers Association (ATDA), and the International Brotherhood of Electrical Workers (IBEW).

Railroad suppliers include a variety of companies that manufacture rail, special components of track such as switches and frogs, signal systems, grade crossing safety devices, communications equipment, rail cars, dispatching software, and locomotives. The Railway Supply Institute (RSI) is the trade association for railroad suppliers. Railroad suppliers often become involved in RSAC working group meetings to help guide discussions of the economic and technical feasibility of proposed regulations. For example, the two major manufacturers of locomotives in the United States, GE Transportation and Electro-Motive Diesel, participated in RSAC meetings concerning working conditions (noise, temperature, vibration, and sanitation) in the locomotive cab (see Chapter 8).

The interplay of FRA with the carriers, labor unions, and suppliers is almost always adversarial. Each group, including the FRA, has an agenda which serves the interests of that group. FRA research is critically examined by the industry stakeholders and challenged, often by experts hired by the stakeholders. Stakeholder input to research design and execution is often sought, but not always forthcoming. There is little or no Human Factors research conducted in the railroad industry other than that by FRA. Consequently, FRA has the task of producing Human Factors research which is relevant to the safety issues of the industry, technically sound, and impartial.

NOTE

1 In the context of grade crossings, there are three distinct groups of individuals who are of concern: railroad personnel, motorists, and trespassers.

REFERENCES

Association of American Railroads. (2025). *Railroad facts* (2021 edn). Washington, DC: Association of American Railroads.

Cothen, G. C., Schulte, C. J., Horn, J. D., & Tyrell, D. C. (2005). Consensus rulemaking at the Federal Railroad Administration. *TR News, 236*, 8–14.

Federal Railroad Administration. (2001). *Rules and roles for FRA staff on RSAC projects.* Unpublished Manuscript.

Federal Railroad Administration. (July 11, 2011). *Federal Railroad Administration Organization Manual*, FRA Order 1100.23e. Washington, DC: Department of Transportation.

Leveson, N. G. (2011). *Engineering a safer world: Systems thinking applied to safety.* Cambridge, MA: MIT Press.

Moray, N. (1994). Error reduction as a systems problem. In M. S. Bogner (Ed.), *Human error in medicine* (pp. 67–91). Hillsdale, NJ: Lawrence Erlbaum.

Moray, N. (2006). Culturing safety for railroads. *Transportation research circular E-C085, Railroad operational safety: status and research needs*. Washington, DC: Transportation Research Board.

Raslear, T. G., & Brennan, C. A. (2013). Fatigue research improves regulatory effectiveness. *TR News, 286*, 37.

Sussman, E. D., & Raslear, T. G. (2008). Railroad human factors. In D. A. Boehm-Davis (Ed.), *Reviews of human factors and ergonomics* (Vol. 3, pp. 148–189). Santa Monica, CA: Human Factors and Ergonomics Society.

2 Social, Legal, and Policy Context

As noted in Chapter 1, the social, legal, and policy context within and outside the railroad industry influences organizational leadership and decision-making, creating the outermost layer of a sociotechnical system from which organizational safety culture and climate flows. Every other element of the sociotechnical system is influenced by this outer layer of the system, and therefore it is important to briefly discuss these influences.

THE SOCIAL CONTEXT

Society constructs expectations about its institutions. The railroad industry and the roles it plays in American society, commerce, and individual lives influence the legal and regulatory constraints that are placed on the industry and the policies that are formulated to promote and control the industry. A concise history of railroads in the United States can be found in *American Railroads* by Stover (1997). Stover outlines the social, legal, and regulatory context that the railroads influenced and were influenced by. To quote the Editor's Foreword to the First Edition by D. J. Boorstin: "…railroads have played a decisive role in nearly every major movement in our history. They were important in hastening the rise of the Atlantic seaboard metropolises, in peopling and supplying the West, in attracting and transporting immigrants, in shaping the enterprises of trappers, cowboys, miners, and farmers. In the twentieth century the commuting train has helped create and perpetuate the American suburbs. Railroads helped determine the course and the outcome of the Civil War and were essential to victory in two world wars" (Stover, 1997, p. xiv).

Because the history of the United States has been so tightly entwined with the development of the railroads, there exists a national sense of connection with railroads. This can be seen in the numerous popular and folk songs about railroads including "The City of New Orleans," "Chattanooga Choo-choo," "John Henry," and "Midnight Train to Georgia." Railroad-themed music exists in jazz, classical, folk, country, world, rock, world, avant-garde, and blues genres (Wikipedia, n.d.). Wikipedia lists nearly 800 titles. Railroads are an American institution.

During my tenure as a research psychologist at the Federal Railroad Administration (FRA), I was surprised to find that many people I interacted with in a non-railroad environment had personal railroad connections. For example, I attended a function at Brown University to honor my dissertation advisor, Dr. Rosemary Pierrel. During the function I spoke with the Psychology Department chairman, Dr. Julius Kling, who asked what I was doing professionally at the time. When I mentioned my research activities at FRA, he became very interested and revealed that he had a railroad connection and wanted to see some of the research reports. I later learned

DOI: 10.1201/9781003500018-3

that he had worked on a railroad before commencing graduate studies in psychology (Providence Journal, 2021).

Part of the reason for the wide personal connection with railroads is that railroading, at least in the twentieth century, was often a family affair (Gamst, 1980). Gamst (1980, p. 30) conducted a survey in which he asked 267 locomotive engineers on three railroads "Did a relative on this railroad in any way help or influence you in becoming an engineer?" Thirty-six percent of the survey sample responded "yes." Rail labor leaders have often said that family connections were important for recruitment of new railroad employees. Gamst provided no contrasting data on other occupations, so it is hard to determine if this is unique to railroading. However, since the 1980s fewer young people are being attracted to railroading. Figure 2.1 shows the mean age of locomotive engineers as a function of years from 1995 to 2021. A linear function fitted to the data accounts for more than 86% of the variance. Anecdotally, railroad management and labor have said that recruiting difficulties may be due to the working conditions that exist in the locomotive engineer occupation, especially those relating to hours of service (HOS) (Kosanda, 2022; also see Chapter 7, Fatigue).

Another factor may be the change in perceptions of the American public concerning railroads and railroading. In the 1800s, when the United States was expanding west, railroads were viewed highly positively (Stover, 1997). The location of railroads brought economic development, population centers, and the transport of farm goods to markets. As railroads gained more economic clout, it was recognized by farmers, in particular, that the railroads were hampering the economic development of others. "The farmer … hated to be at the mercy of the local elevator company, often owned by the railroad and managed by a man who downgraded the farmer's grain and who could frequently afford to hold it until a profitable price rise came along. Nor did the farmer like the fancy prices he paid at the local country store, priced high in part because the merchant was in the same freight-bind as the farmer" (Stover, 1997, p. 115). The public perception of railroads became more negative. Popular literature reflected this in novels

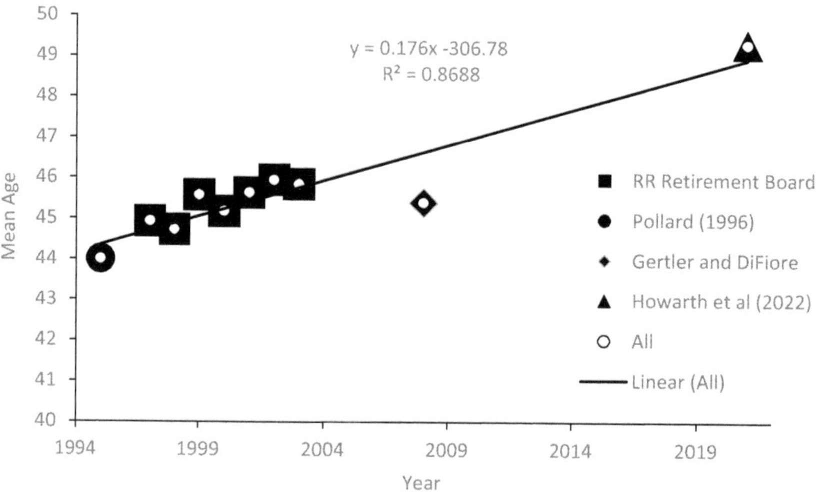

FIGURE 2.1 Mean age of locomotive engineers as a function of year.

such as *The Octopus* by Frank Norris (1901/1994). The railroads were monopolies and acted accordingly. They became corrupt and incompetent. As the public soured on the railroads, the economic philosophy and mindset of the United States swung from laissez faire to an increase of governmental regulation of industry in general.

THE LEGAL CONTEXT

For a complete list of important dates in railroad history see Stover (1997, pp. 263–272). Here we will focus on the most important legislative actions that provide the context in which Human Factors research in the railroad industry occurs.

In 1872, the United States Senate Committee on Transportation Routes to the Seaboard, also known as the Windom Committee, was authorized to examine transportation between the inland regions of the United States and the coast (Windom, 1881; 2018). The committee produced a report to Congress in 1874 that called for cheap and efficient transportation of goods between the interior and seaboard areas of the United States. The committee called for the Congress to prescribe legislation to accomplish that goal. Questions were raised concerning the ability of the states to regulate interstate commerce under the Commerce Clause of the Constitution, thereby setting the stage for extension of congressional authority into this area. State legislation (see Granger laws, Stover, p. 120 ff.) regulated the railroads, and the railroads launched legal challenges to that legislation. In 1876 the Supreme Court made several decisions that were favorable to the Granger laws. But in 1886 the Supreme Court in the Wabash Case ruled that a state could not regulate shipment rates beyond its borders. This set the stage for the passage by Congress in 1887 of the Interstate Commerce Act. The Act required that all interstate rates be reasonable and just. It prohibited several practices such as rebates, drawbacks, and pools as well as long-haul and short-haul rate discrimination. It also required that railroads publish shipping rates. It created the Interstate Commerce Commission (ICC) to monitor the railroads regarding compliance with the legislation. Consequently, the railroads became the first industry to be directly regulated by the U.S. government (Wentworth, 1891/2018). The ICC was the forerunner of the FRA regarding safety regulations and oversight. The FRA was created by the Department of Transportation Act of 1966. The ICC Termination Act of 1995 abolished the ICC and transferred many of its remaining functions to the Surface Transportation Board (STB) in the Department of Transportation. "The Surface Transportation Board is an independent federal agency that is charged with the economic regulation of various modes of surface transportation, primarily freight rail. The agency has jurisdiction over railroad rate, practice, and service issues and rail restructuring transactions, including mergers, line sales, line construction, and line abandonments. The STB also has jurisdiction over certain passenger rail matters, the intercity bus industry, non-energy pipelines, household goods carriers' tariffs, and rate regulation of non-contiguous domestic water transportation (marine freight shipping involving the mainland United States, Hawaii, Alaska, Puerto Rico, and other U.S. territories and possessions)" (Safety Transportation Board, n.d.). Today's regulation of railroad economic and safety issues lies with the STB and FRA.

The 1970s and 1980s saw an increase in Congress' deregulation of the railroads. The most important of these legislative actions was the Staggers Rail Act of 1980

(United States Congress, 1981/2018; Association of American Railroads, 2022). During this time period railroads regained the ability to set freight rates, to eliminate unprofitable routes, and to determine what services to offer. Unprofitable routes were sold to the new short lines which offered local and regional rail freight services. According to the Association of American Railroads (2022), "Since Staggers was passed, average rail rates have fallen 44%, safety has dramatically improved, rail traffic volume is up 90% and railroads have poured nearly $740 billion—their own funds, not taxpayer funds—back into their networks." Prior to the passage of the Staggers Act many railroads were facing bankruptcy or were bankrupt. Deregulation of the railroads allowed them to be competitive with the trucking industry, to increase productivity, and to thrive economically.

The experience of the railroads with stringent regulation and economic decline during the period prior to 1980 and the subsequent post-1980 revival under deregulation has fostered a fierce opposition by the railroads to all and any regulation of their activities and a cautious attitude toward any federal involvement (financial or technical) in their affairs. This view of the regulatory, technical, and economic functions of the FRA has often resulted in difficult negotiations to act cooperatively on projects that would benefit both the public and the railroads' interests. A primary mission of the FRA is the safety of rail operations. FRA projects that promote safety often result in new *regulations* because compliance with safety *recommendations* is often poor. Because new regulations impose costs on the railroads (and also often on rail labor), new regulations are often opposed by the carriers and rail labor. However, not all safety research has to result in new regulations. The safety culture projects that FRA's Human Factors Division fostered (see Chapter 9) were not undertaken with an intention to formulate new regulations. Yet it was extremely difficult to obtain industry participation in such projects. It is in this sense that the legal context colors all of the other elements of the sociotechnical system. To be fair, this is not only true of the interactions of railroad management with the FRA but also that of railroad labor and railroad suppliers. A bankrupt railroad does not employ workers or buys equipment and materials. The statement "I am from the Federal government, and I am here to help you," consequently, is anathema for all. Doing research in the railroad environment involves negotiating conflicting interests of several entities.

Another major piece of legislation that sets the context for Human Factors research in the railroad industry is the Federal Employers' Liability Act (FELA; Committee for Study of the Federal Employers' Liability Act, 1994; Reinach and Gertler, 2001, pp. 18–22). FELA is a compensation system for on-the-job injuries of railroad workers. FELA was passed in 1908, the same year that Congress also established a workers compensation system for federal employees. The federal employee system is widely used in the United States to cover most other workers. Each state has its own version of this system. The two systems differ in several ways, the most important of which is that current workers' compensation systems are no-fault, while FELA is tort-based. FELA allows railroad workers "… to sue their employers for all damages related to on-the-job injuries without limit or restriction, including compensation for medical expenses, lost wages, disabilities, future earnings losses, and pain and suffering. To collect, the injured worker has to demonstrate negligence on the part of the employer, and awards can be reduced depending on the degree of employee negligence through a 'comparative negligence' standard. An injured worker can receive no compensation at all if the injury is judged to be entirely the worker's fault" (Committee for Study of the Federal Employers' Liability Act, 1994, p. 2). In contrast with the states' worker compensation systems, FELA provides a higher and more extensive level of benefits. The railroads, in part, would prefer an alternative compensation system because of the higher costs FELA entails. The railroads also argue that these higher costs reduce their competitive posture relative to other modes of transportation that are not subject to FELA.

The primary mission of the FRA is railroad safety. The bulk of Human Factors research at FRA is aimed at improving railroad safety. Rail labor advocates for FELA on the basis that it provides a strong incentive to improve the safety of their system. If the railroad is found to be negligent, accountability in the form of compensation to the injured worker provides an incentive for the railroad to remedy the cause of the injury. Rail management points to the improvements in railroad safety subsequent to deregulation after the passage of the Staggers Act (see above). Rail management argues that the cost of accidents with regard to freight, equipment, and infrastructure damage is a stronger incentive for accident avoidance. Tellingly, "Railroad management further contends that the fault-based nature of the FELA system fosters an adversarial relationship between management and labor. This impedes the ability of both parties to fully understand and correct unsafe workplace practices. Full investigation of accidents becomes difficult because the incentives for each part are not to cooperate in revealing any facts or circumstances that would indicate their fault. This undermines the working relationship with employees and hinders efforts to improve the safety of the workplace environment" (Committee for Study of the Federal Employers' Liability Act, 1994, p. 22). Although the investigation of railroad accidents is the purview of the FRA Office of Safety, the Human Factor Research Division of the Office of Research and Development does research to aid and improve accident investigation and understanding. FELA provides part of the context in which that research is conducted. Accidents often involve human error, which can be due to working conditions, physiological issues, equipment design, weather, and/or operational procedures. Understanding human error causes in accidents and near accidents would enhance safety, but is often hampered by the reluctance of rail labor and management to cooperate in research. FELA is an important context for Human Factors research in the railroad environment (also see Reinach and Gertler, 2001, p. 21).

THE POLICY CONTEXT

Policy is generally set by the Congress through legislation (the strategy) and implemented by government agencies through regulations (the tactics). Congress decides through appropriation and authorization bills what each agency can and cannot do. When signed into law by the President, these bills are known as statutes and are codified in the U.S. Code as Public Laws (PL). "A typical statute may establish an obligation or a mandate on behalf of a Federal agency. In order to comply with that mandate, the agency will promulgate a regulation as to how the agency will enforce compliance with the statutory requirements" (U.S. Department of the Interior, n.d.). The regulations are codified in the Code of Federal Regulations (CFR).

For instance, FRA for many years sought the ability to directly regulate HOS for various railroad employees (see Chapter 7, Fatigue). Rather, Congress dictated the specific parameters of the regulations until the passage of the Rail Safety Improvement Act of 2008. In 1989 Congress amended The Federal Hours of Service Act of 1907 (49 United States Code 21101-21108) to prescribe the length of on-duty and off-duty time for certain railroad employees. This provided the legal basis for FRA's HOS regulations (49 Code of Federal Regulations Part 228). Consequently, FRA actions to improve safety in the railroad industry were limited to enforcement of HOS regulations that Congress enacted in law. Research published in 1997 (Thomas et al., 1997) had indicated that those regulations did not prevent fatigue, but policy limited FRA regulatory action for more than a decade. Research, however, continued on the fatigue issue in anticipation of Congress relenting and partially allowing FRA to directly decide what regulatory actions were best suited to reduce fatigue in the industry. In 2008 Congress passed the Railroad Safety Improvement Act of 2008 (RSIA; 49 United States Code 20101), which gave FRA the authority to regulate HOS in general, and specifically to determine HOS suitable for passenger and commuter rail service train crews. If the research had not continued in advance, deadlines for action set by Congress under RSIA might not have been met. If the agency had not recognized the need for direct regulatory control of HOS, such research might not have been prioritized within FRA.

Human Factors research agendas at FRA were often set by the needs of the FRA's Office of Safety for support of its regulatory agenda. Since much policy from Congress in the form of appropriation and authorization legislation was aimed in a general way toward the improvement of railroad safety (as documented in FRA accident and incident data), the Office of Research and Development (R&D) and the Office of Safety often relied on observations of the industry's safety performance (changes in accidents and incidents), complaints from railroad labor, and catastrophic accidents to direct new research and possible regulatory actions. Catastrophic accidents, which garner public attention in news media and, consequently, the attention of Congress, are particularly poignant events for a regulatory agency. For instance, the Lac Megantic accident (Transportation Safety Board of Canada, 2014) resulted in a comprehensive effort by R&D to demonstrate the relevance of Human Factors, Equipment and Rolling Stock, Track, Signal, and Hazmat research (all separate research divisions in R&D) on that accident (Raslear, 2013; Tunna, 2014). This is a good example of how the social context influences legal and policy contexts, and, in turn, research. Fortunately, FRA R&D had already identified a number of safety issues for research that were relevant to the accident.

Railroads as an institution provide another context for the conduct of research by the federal government. Railroad management considers research to be regulatory *per se*. When a government research report is published, the published report can become part of the evidence in a FELA tort to establish negligence by the railroad. For instance, FRA published a report concerning safety in rail yards (Reinach and Gertler, 2001). They found that "… slips, trips and falls account for the largest number of injuries (42 percent) as well as the largest proportion of lost workdays (44 percent)" (p. 49). Among causes of slips, trips, and falls, the highest proportion involved ballast and other objects on the yard surface (p. 50). Ballast alone accounted for the second highest number of injuries involving roadbed materials (p. 60, Table 25). In focus groups with yardmasters, "Distribution of smaller 'walking' stone on switch leads instead of the larger ballast rock" was identified as one of the "factors that might help reduce railroad yard injuries" (p. 105, Table 43). This report (personal communication) has been cited in FELA cases in which railroad workers suffered injuries from slips, trips, and falls. Apparently, government reports are treated as fact in courts and railroads can be considered negligent if they do not follow recommendations in those reports to improve safety. (See Chapter 8, Working Conditions and Ergonomics, for more details about the ergonomics of ballast.)

Government agencies are sometimes considered the fourth branch of government (Smith and Licari, 2006). This is because agencies such as the FRA engage in rulemaking. "Rulemaking... is a legislative function that agencies use to make their own policy. Rules made by government agencies carry the force of law, and are used to both implement and clarify public policy made by Congress. Rulemaking allows agencies to use their expertise, information, and discretion to codify how they will achieve the broader policy goals set forth by Congress" (Smith and Licari, 2006, p. 89).

In fact, rulemaking by the agencies and lawmaking by the Congress are often reciprocal activities. Agencies have specialized experts on subjects within their domain and continuously gather information from formal and informal sources about those subjects. As noted above, FRA expertise and information is generally used to direct new research and possible regulatory actions. This expertise and information is also often used by agencies to draft legislation that is proposed to Congress or specifically requested by a congressional committee or member. This has certainly been the case at FRA.

Laws passed by Congress often are vague or lack specifics concerning the application of the policy goal of the law. Consequently, agencies draft regulations that supply their interpretation of the law and set the specifics of the policy. The rulemaking process is itself strictly regulated by the Administrative Procedures Act of 1946 (Smith and Licari, p. 89). The Act requires that a draft of the proposed rule be submitted to the Office of Management and Budget (OMB) for review, be published in the Federal Register for comment by the public and interested parties, and then either be finalized (with or without changes suggested by the comments) or abandoned. The process of drafting a rule can be accomplished in various ways. FRA has traditionally used formal negotiated rulemaking. More recently, consensus rulemaking, which employs a formally constituted Advisory Committee, has been used extensively (see Cothen et al., 2005; Federal Railroad Administration, 2001 for a detailed description of each process). In both approaches, interested parties and stakeholders are involved in formulating the rule. The final draft of the rule is written by FRA legal staff. Constraints involving the Paperwork Reduction Act, economic and environmental impact, etc., are also applied. Before a rule is presented to OMB it is reviewed by the FRA Administrator and the Office of the Secretary of Transportation.

As has been noted in Chapter 1 and above, agency research is also involved in the rulemaking process and, consequently, in policy making. Advisory committees are often constituted in part or wholly as scientific advisory committees. Throughout the federal government, such advisory committees are so widely used that some consider advisory committees to be the fifth branch of government (Jasanoff, 1990). At FRA, consensus rulemaking takes place within the context of the Railroad Safety Advisory Committee (RSAC).

Advisory committees provide agency officials with knowledgeable and up-to-date experts in highly specialized areas. Within the RSAC at FRA, agency experts and scientists serve this role, particularly with regard to Human Factors. Research reports are often cited in FRA rules to support the rulemaking, and, as noted above, research reports can act as rules *per se* if used in legal actions under FELA to establish railroad negligence. For this reason railroad management has accused Human Factors project managers of regulating by research. Yet, as Jasanoff notes, regulatory actions without expert advisors are prone to legal and political challenges which target the quality and sufficiency of technical arguments. So, agencies need the expertise, and the regulated entities consider the expertise to be unsanctioned regulation. As Jasanoff (p. 9) succinctly puts it, "… if regulatory advisors invariably become part of a hybrid sociotechnical process, as most of the literature suggests, then how can they maintain their authority as neutral experts, especially when challenged in the media or the courts?"

Needless to say, many regulatory actions by FRA have been challenged in court. But because the Class I railroads stopped Human Factors research in the 1990s at their Chicago AAR facilities, they have had little recourse to their own research evidence. This has led the Class I railroads to resort to hiring academic experts to produce research that contradicts FRA research or to contradict it in reviews. This is a tactic widely employed by other industries such as the tobacco and petroleum industries (Oreskes and Conway, 2010). This adversarial approach slows regulatory action and discourages participation of independent researchers in the regulatory process.

As an example, the FRA has consistently received complaints from railroad workers about back problems that result from walking on ballast. An independent researcher with considerable expertise on this topic proposed an "… extensive study on the same general topic … to all of the Class 1 railroads, the railroad trade association (Association of American Railroads, AAR) and the Federal Railroad Administration (FRA). The FRA eventually funded the study, but not before railroad defense attorneys contacted the General Counsel at the FRA and attempted to prevent the study from being funded. The AAR then tried to insert adversarial experts of its choosing into any peer-review done of the study by FRA, at which point the FRA and investigators decided to go directly to a refereed journal with any manuscripts instead of providing a report to the funding agency. … Railroad defense attorneys issued more than 30 subpoenas to co-authors and their institutions to obtain the original manuscript and reviews" (Andres, 2011). The research was published (Wade et al., 2010), but the experience serves "… as a cautionary tale for experts who also find themselves involved in original research" (Andres, 2011). Potential legal entanglements could cause independent researchers to avoid undertaking contract research with a federal agency. Oreskes and Conway (2010, pp. 262–265) document the various other factors which intimidate scientists and other experts from engaging in activities that expose them to legal, political, and peer scrutiny.

Although the publication of contradictory data is usually associated with industry attempts to discredit federal agency research, this can also happen within an agency when other agency departments perceive a rivalry with the agency's research department and individuals' perceived status is threatened. The research surrounding this incident at FRA is discussed in detail in Chapter 4, Exposure. Herewith is a brief summary.

Members of the railroad industry and the FRA's Office of Railroad Safety requested help from the FRA's Human Factors Research Division in interpreting time of day accident data involving maintenance of way (MOW) workers and signalmen. The original FRA report, *"Is There a Time of Day Effect for Maintenance of Way Worker and Signalman Fatalities? Why Risk Exposure is Important"* (Kidda and Raslear, 2014), was not published because of objections by a member of the FRA's Office of Railroad Safety. Kidda and Raslear (2014) concluded that fatality risk was greatest between the hours of 2000 and 0300, which was at odds with the conclusion favored by the member of the FRA's Office of Railroad Safety that fatality risk was greatest at 1300h. Research was consequently funded by FRA's Human Factors Research Division to examine how time of day affects injury risk (providing a much larger and robust sample size for statistical analysis than fatalities) of railroad MOW employees and signalmen. The research was published in the *Journal of Safety Research* (Calabrese et al. 2017) and agreed with the conclusion of Kidda and Raslear (2014) concerning risk. Two years later, the FRA published a report by Kumagai and Harnett (2019). Their report showed that the greatest risk of accidents and incidents occurred at 1200h.

However, close scrutiny of the report raises concerns as noted by Raslear (September 2, 2019). In his letter to Kumagai and Harnett, he notes that "… non-identical data is included in both figures for 0:00h. For example, in Fig. 2 [of Kumagai and Harnett] the left abscissa data for 0:00 is 1% (5 cases) and the right abscissa data for 0:00 is 3% (18 cases)." This is shown in Figure 2.2. Note that in this figure there are 25h in a day rather than the expected 24. Raslear in his letter also asked for information about how the data were normalized in Kumagai and Harnett's Figure 2, shown here as Figure 2.3. The data appear to be Kumagai and Harnett's.

There is no description in Kumagai and Harnett (2019) of how the data were normalized. Furthermore, the letter to Kumagai and Harnett was unanswered. It was returned by the Post Office with a note indicating that the addressees were unknown at the address provided for correspondence in the report. Given that the Kumagai and Harnett data support the original claim by the member of the Office of Railroad Safety, it is easy to suspect that the Kumagai and Harnett report is an attempt to cast doubt on the data in the Calabresse et al. (2017) report.

OTHER ENTITIES THAT INFLUENCE POLICY

Government Accountability Office

In addition to organizations within the railroad industry and the FRA, the U.S. Government Accountability Office[1] (GAO) plays an important role in policy making. "GAO provides Congress, the heads of executive agencies, and the public with

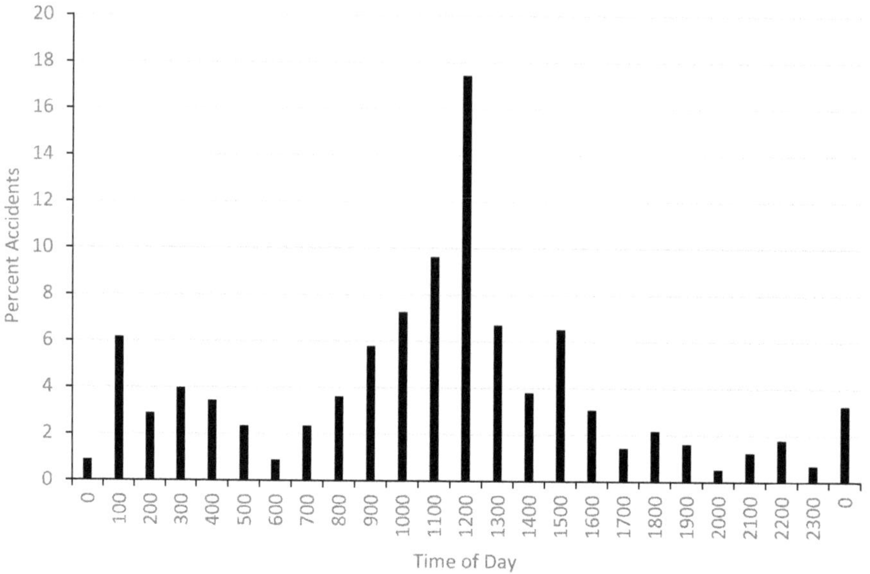

FIGURE 2.2   Percent of accidents as a function of the 24-hour day. (After Kumagai and Harnett 2019.)

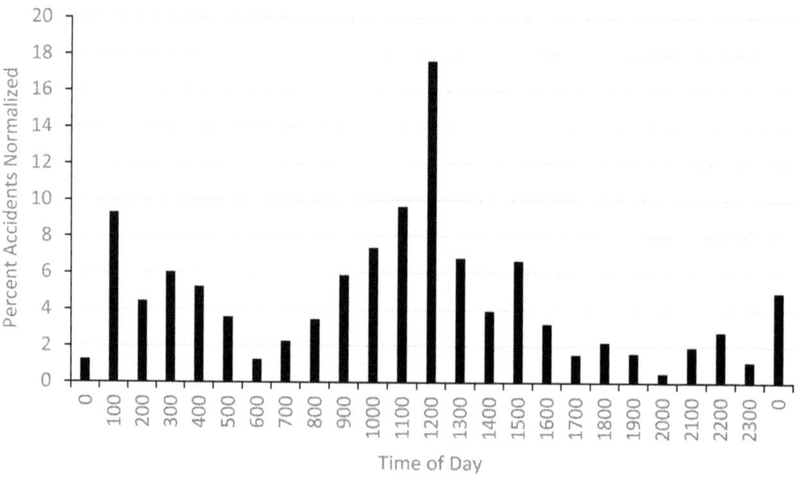

FIGURE 2.3   Percent of accidents as a function of the 24-hour day (normalized for exposure). (After Kumagai and Harnett 2019.)

timely, fact-based, non-partisan information that can be used to improve government and save taxpayers billions of dollars" (United States Government Accountability Office, 2022). Congress, from time to time, requests that GAO provide a report or other information about the status of the railroad industry or the activities of the FRA. For example, Congress requested that the GAO examine the issue of fatigue and work schedules in the railroad industry in the 1990s. GAO produced a report and a briefing to Congress (United States General Accounting Office, 1992, 1993) which supported FRA's claim that the 1907 Hours of Service Act should be repealed in favor of regulations that would allow a more flexible approach to HOS and scheduling. When such changes were enacted by Congress in 2008 in the RSIA, Congress consequently requested that GAO examine the outcome of the changes implemented by FRA through regulations (United States Government Accountability Office, 2011; United States Department of Transportation, 2012; also see Chapter 7, Fatigue; Raslear, 2019). Reports from the GAO are important for informing Congress about issues affecting policy and funding, and for FRA in understanding what it may or may not have gotten right in regulating the railroad industry. However, the relationship with FRA is often adversarial, and the conclusions reached by GAO are sometimes erroneous because the highly technical issues that GAO examines are staffed by individuals who lack sufficient expertise in the topic (Raslear, 2019). Although GAO staff request expert help from FRA, the advice of FRA staff is sometimes not heeded.

In addition to formal institutions within the executive, congressional, and judicial branches of government, there are other independent organizations which also exert substantial policy influence on FRA. These include the Transportation Research Board (TRB) which is part of the National Academies of Science and the National Transportation Safety Board (NTSB).

Transportation Research Board

The role of TRB in policy making for the FRA is complex. As an independent organization, TRB affects policy by informing the engineering and scientific communities about issues of concern in the railroad industry through its publications and annual meeting. FRA has taken advantage of that TRB role by actively participating in the annual meeting (presenting papers, staffing a booth in the exhibition area, attending meetings of standing committees, participating in the planning committee for Human Factors Workshops) and by helping to found the TRB Standing Committee on Railroad Operational Safety (AR070). Through these activities, FRA establishes research needs and promotes its accomplishments within an international forum.

Another role of TRB is "… to convene a committee of experts to review the quality and relevance of RD&T's [FRA's Office of Research, Development and Technology] current and planned research portfolio and to provide advice on strategies to better identify research needs, conduct high-quality research, and ensure that research products contribute to FRA's primary goal of improving railroad safety" (Transportation Research Board, 2022). The Committee for Review of the FRA Research, Development, and Demonstration Programs was initiated in 1998. The committee issues reports which can be found on the TRB and FRA websites. The 2020 report of the committee is typical in that it provides an honest, independent overview of the R&D program by experts (Transportation Research Board, 2020). Committee recommendations regarding funding and research priorities are important factors that influence research policy at FRA.

National Transportation Safety Board

The NTSB plays a very important role in transportation policy. "In 1974, Congress reestablished the NTSB as a completely separate entity, outside the DOT, reasoning that '... No federal agency can properly perform such (investigatory) functions unless it is totally separate and independent from any other ... agency of the United States.' Because the DOT has broad operational and regulatory responsibilities that affect the safety, adequacy, and efficiency of the transportation system, and transportation accidents may suggest deficiencies in that system, the NTSB's independence was deemed necessary for proper oversight. The NTSB, which has no authority to regulate, fund, or be directly involved in the operation of any mode of transportation, conducts investigations and makes recommendations from an objective viewpoint" (National Transportation Safety Board, 2022). Through accident investigations and recommendations the NTSB exerts a great deal of influence over policy goals for DOT and FRA. Accidents that merit NTSB investigation are very often high profile and receive broad press coverage which ensures the attention of Congress, the public, and safety experts. The NTSB is meticulous in its procedures and has experts who specialize in each mode of transportation and in specific engineering and scientific fields such as metallurgy and Human Factors. For years the NTSB has published a Most Wanted List of Transportation Safety Improvements (National Transportation Safety Board, 2021-2022). When this list included fatigue, it helped provide a rationale for FRA's recommendation that Congress allow it to directly regulate fatigue (see Chapter 7, Fatigue).

NOTE

1 Previously known as the General Accounting Office.

REFERENCES

Andres, R. O. (2011). Litigation and attempts to pierce the peer-review process. In *American Academy of Forensic Sciences Annual Scientific Meeting*, February 21–26, 2011, Chicago, IL.

Association of American Railroads. (2021). *Railroad facts* (2021 edn). Washington, DC: Association of American Railroads.

Association of American Railroads. (2022). *Freight railroads & the Staggers Rail Act of 1980*. Washington, DC: Association of American Railroads.

Calabresse, C., Mejia, B., McInnis, C. A., France, M., Nadler, E., & Raslear, T. (2017). Time of day effects on railroad worker injury risk. *Journal of Safety Research*, *61*, 53–64.

Committee for Study of the Federal Employers' Liability Act. (1994). *Compensating injured railroad workers under the Federal Employers' Liability Act* (Special Report 241). Washington, DC: National Academy Press.

Cothen, G. C., Schulte, C. J., Horn, J. D., & Tyrell, D. C. (2005). Consensus rulemaking at the Federal Railroad Administration. *TR News, 236*, 8–14.

Federal Railroad Administration. (2001). *Rules and roles for FRA staff on RSAC projects*. Unpublished Manuscript.

Gamst, F. C. (1980). *The hoghead. An industrial ethnology of the locomotive engineer*. New York: Holt, Rinehart and Winston.

Gertler, J., & DiFiore, A. (2009). *Work schedules and sleep patterns of railroad train and engine service workers* (Report No. DOT/FRA/ORD-09/22). Washington, DC: U.S. Department of Transportation.

Howarth, H. D., France, M. E., Snow, J. Z., & Harnar, M. A. (2022). *Information and communications technology survey of Class I railroad train, yard, and engine workers* (Report No. DOT/FRA/ORD-22/03). Washington, DC: U.S. Department of Transportation.

Jasanoff, S. (1990). *The fifth branch: Science advisors as policymakers.* Cambridge, MA: Harvard University Press.

Kidda, S., & Raslear, T. G. (2014). I*s there a time of day effect for maintenance of way worker and signalman fatalities? Why risk exposure is important.* Unpublished manuscript.

Kosanda, A. L. (April 22, 2022). Letter to Surface Transportation Board Chairman Marty Oberman*: At What Point is This Not Worth It?* Railway Age.

Kumagai, J. K., & Harnett, M. (2019). *Data analysis for maintenance-of-way worker fatigue* (Report No. DOT/FRA/ORD-19/02). Washington, DC: U.S. Department of Transportation.

National Transportation Safety Board. (2021–2022). *2021–2022 most wanted list.* www.ntsb.gov/advocacy/mwl/Pages/default.aspx (accessed May 18, 2022).

National Transportation Safety Board. (2022). *History of the National Transportation Safety Board.* www.ntsb.gov/about/history/Pages/default.aspx (accessed May 18, 2022).

Norris, F. (1901/1994). *The octopus: A story of California.* New York: Penguin Classics.

Oreskes, N., & Conway, E. M. (2010). *Merchants of doubt. How a handful of scientists obscured the truth on issues from tobacco smoke to global warming.* New York: Bloomsbury Press.

Pollard, J. K. (1996). *Locomotive engineers' activity diaries* (Report No. DOT/FRA/RRP-96/02). Washington, DC: U.S. Department of Transportation.

Providence Journal. (October 21, 2021). *Julius W. "Jake" Kling Obituary.*

Raslear, T. G. (2013). *Chiefs Meeting 08/27/2013.* Unpublished manuscript.

Raslear, T. G. (2019a). *Six safety letter 2Sept19.docx.* Unpublished letter in reference to Kumagai, J. K., & Harnett, M., 2019 (DOT/FRA/ORD-19/02).

Raslear, T. G. (2019b). *The GAO reports on hours of service, 1992-1993.* Unpublished manuscript.

Reinach, S., & Gertler, J. (2001). *An examination of railroad yard worker safety* (Report No. DOT/FRA/ORD-01-20). Washington, DC: U. S. Department of Transportation.

Safety Transportation Board (n.d.). *About STB.* stb.gov (accessed February 9, 2024).

Smith, K. B., & Licari, M. J. (2006). *Public administration. Power and politics in the fourth branch of government.* Los Angeles, CA: Roxbury Publishing Company.

Stover, J. F. (1997). *American railroads* (2nd edn). Chicago, IL: University of Chicago Press.

Thomas, G. R., Raslear, T. G., & Kuehn, G. I. (1997). *The effects of work schedule on train handling performance and sleep of locomotive engineers: A simulator study* (Report No. DOT/FRA/ORD-97/09). Washington, DC: U.S. Department of Transportation.

Transportation Research Board. (2020). *Review of the Federal Railroad Administration's research and development program.* Washington, DC: National Academies Press.

Transportation Research Board. (2022). *Review of the Federal Railroad Administration's research and development program.* Blurbs. trb.org (accessed May 18, 2022).

Transportation Safety Board of Canada. (2014). *Runaway and main-track derailment, Montreal, Maine & Atlantic Railway, freight train MMA-002, mile 0.23, Sherbrooke subdivision Lac-Mégantic,* Quebec, 6 July 2013 (Railway Investigation Report R13D0054). Gatineau, QC: Transportation Safety Board of Canada.

Tunna, J. (2014). *RPD_30_All_Hands_February2014_MI_MEK_Edits* (2020_11_06 22_03_58 UTC). Unpublished PowerPoint presentation.

United States Congress. (1981/2018). *Staggers rail act of 1980* (Classic reprint). London: Forgotten Press.

United States Department of the Interior. (n.d.). *Federal statutes and regulations.* doi.gov. (accessed March 29, 2022).

United States Department of Transportation. (2012). *Letter to Congress* (Report No GAO-11-853). Washington, DC: United States Department of Transportation.

United States General Accounting Office. (1992). *Railroad safety: Engineer work shift length and schedule variability* (Report No. GAO/RCED-92-133). Washington, DC: United States General Accounting Office.

United States General Accounting Office. (1993). *Railroad safety: Human factor accidents and issues affecting engineer work schedules* (GAO/RCED-93-160BR). Washington, DC: United States General Accounting Office.

United States Government Accountability Office. (2022). *What GAO does*. GAO.Gov (accessed May 16, 2022).

Wade, C., Redfern, M. S., Andres, R. O., & Breloff, S. P. (2010). Joint kinetics and muscle activity while walking on ballast. *Human Factors: The Journal of the Human Factors and Ergonomics Society, 52,* 560–573.

Wentworth, J. T. (1891/2018). *The interstate commerce law: being the act to regulate commerce, which was approved February 4, 1887, and in effect* April 5, *1887; as amended by act ... in effect February 10, 1891* (Classic Reprint). London: Forgotten Press.

Wikipedia. (n.d.). *List of train songs* (accessed March 15, 2022).

Windom, W. (1881/2018). *Report of the select committee on transportation-routes to the seaboard: 1.* Sacramento, CA: Palala Press.

Section II

Analytic Methods

3 Data Issues

Except when a research project directly collects information about a safety issue in the railroad industry, the data used to characterize the safety of the industry come from information that the railroad industry is required by law to provide to the Federal Railroad Administration (FRA). U.S. railroads are required to provide FRA with accident/incident reports. The FRA collects this data and makes it available to the public on the FRA website (Federal Railroad Administration, n.d.) in databases. Events that must be reported include train accidents (collisions, derailments, and other events that involve the operation of on-track equipment that causes damage above a threshold that is adjusted over time), highway-rail grade crossing incidents, and other incidents that cause a fatality or an injury to any person, or an occupational injury to a railroad employee. Each reportable event is coded with regard to the circumstances of the event (e.g., cause, location, time of day, action being performed, etc.). An excellent detailed description of the FRA data and how it can be used in the analysis of a safety issue can be found in Reinach and Gertler (2001).

The FRA has cause codes that relate reportable events to major categories such as Human Factors, Signal Defects, Track Defects, Equipment/Mechanical problems, and Miscellaneous. I have often heard it said that cause codes are assigned arbitrarily to events, or that sometimes codes are assigned to benefit the railroad department that arrives first at the scene (e.g., a track defect is assigned a mechanical code by a track department employee). As with any reporting database, users must be concerned about the validity of the data and the integrity of the reporting system (see United States Government Accountability Office, 2018).

HOW ARE CAUSE CODES ASSIGNED?[1]

The recording of the causation of events, such as accidents and deaths, has a long history of error and bias. Death certificates have long been known to be susceptible to errors, and standard textbooks of epidemiology (Lilienfeld and Lilienfeld, 1980; Gordis, 2014) always discuss this issue. For instance, "For the major cardio-vascular-renal diseases, only 32.8 percent of deaths so certified were solidly established as compared to 89.3 percent for malignant neoplasms of the lymphatic and hematopoietic tissues. However, at the other end of the scale, the percent of cause-of-death statements that were considered to be probably wrong varied from 0 to 20.3 percent" (Lilienfeld and Lilienfeld, 1980, pp. 70–71).

Among the FRA cause codes, there are 118 human factor codes out of a total of 389 cause codes. This means that 30.339% of all cause codes are human factors. Interestingly, the historic percentage of human factors accidents ranges from 25% to 41% with an average of 35% (1978–2017). Are there biases in how accident cause codes are assigned, and, if so, how can those biases be addressed?

DOI: 10.1201/9781003500018-5

Table 3.1 shows the number and proportion of cause codes for the major categories of causes.

Figure 3.1 shows the 15-year trend (1990–2004) in the proportion of accidents for each major cause code. The lines are trend lines for each category and show that only signal codes show a reliable increase over the 15-year period. Some of the accident proportions in Figure 3.1 approximate the proportion of cause codes expected from code proportion. Human factors has 0.3669, miscellaneous has 0.1367, and signal has 0.0153 of the accidents. However, track has 0.3513 of accidents, which exceeds its cause code proportion and mechanical has 0.1298 of accidents, which is below its cause code proportion. To further evaluate the extent to which cause code proportion could account for the number of accidents attributed to various causes, the expected frequency of accidents attributed to each cause code in each year was calculated and compared with the observed frequency of accidents. For the 15 years of data and 5 cause code categories this resulted in 75 data points. This is plotted in Figure 3.2

TABLE 3.1

Frequency and Proportion of the Major Cause Codes

Cause	Frequency	Proportion
Track	65	0.1671
Signal	21	0.054
Equipment	143	0.3676
Human Factors	118	0.3033
Miscellaneous	42	0.108
Total	389	1.0

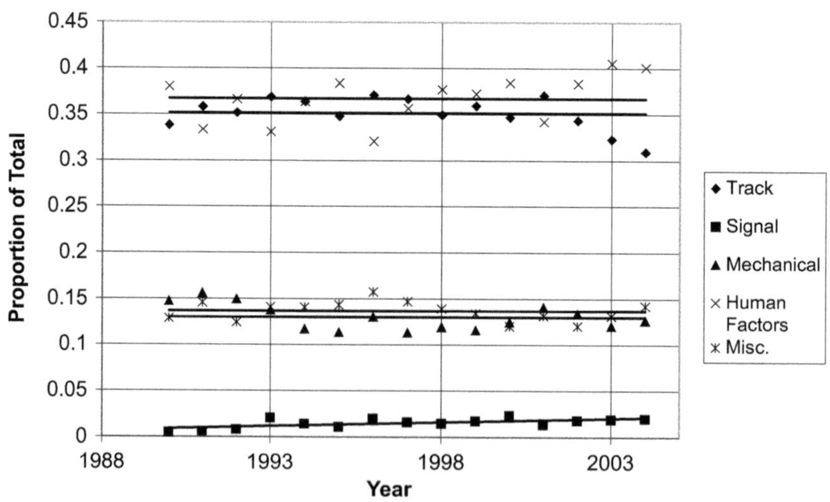

FIGURE 3.1 Proportion of total accidents for each major cause code.

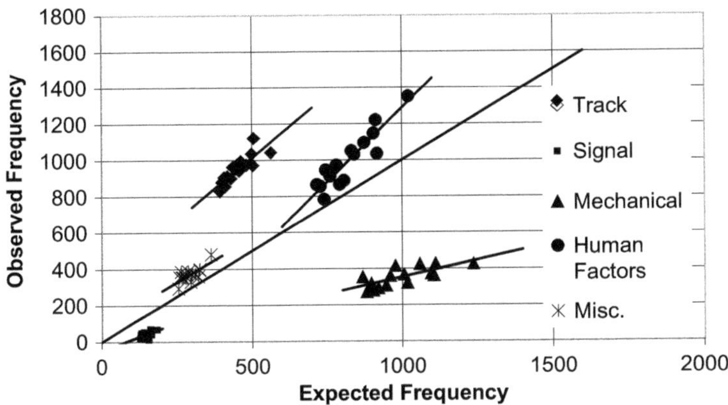

FIGURE 3.2 Observed frequency of accidents in each cause code by expected frequency.

by cause code. The large solid diagonal line in Figure 3.2 is the line the data should fall on if the frequency of accidents in each cause code was perfectly predicted by the proportion of cause codes. As noted before, there is not a perfect prediction by proportion of cause codes, but there is some indication that the proportion of cause codes influences the assignment of accident causes. To assess the size of this effect, a correlation between the 75 expected and observed frequencies was conducted. The correlation coefficient was 0.4281. This is statistically reliable ($p < 0.01$) and indicates that 18.32% of the variance in observed accident frequencies in each cause code can be accounted for by the available proportion of accident cause codes. However, it should be noted that the number of cause codes in each category might just reflect the number of accident causes in that category.

Another factor that could influence the frequency of accidents reported in each cause code category is the number of inspectors in each discipline. This information is shown in Table 3.2.

Note that there are no miscellaneous inspectors. There are, however, 79 supervisory field personnel including chief inspectors, regional administrators, deputy regional administrators, and supervisory railroad safety specialists. It is safe to assume that these non-discipline-specific personnel focus part of their attention on the miscellaneous causes of accidents as well as the other four categories. Assuming

TABLE 3.2

Number and Proportion of Inspectors

Inspectors	Number	p
Equipment	86	0.290541
Human Factors	76	0.256757
Signal	61	0.206081
Track	73	0.246622
Misc.	0	0
Total	296	1

TABLE 3.3
Revised Number and Proportion (*p*)
of Inspectors

Inspectors	Number	*p*
Equipment	89.95	0.239867
Human Factors	79.95	0.2132
Signal	64.95	0.1732
Track	76.95	0.2052
Misc.	63.2	0.168533
Total	375	1

that they put 5% of their effort into each of the discipline-specific categories and the remainder into the miscellaneous category, we would have the revised number of inspectors shown in Table 3.3.

There is also a reliable correlation ($p < 0.01$) between the expected frequency of cause codes attributed to accidents (based on the revised proportion of inspectors in each discipline) and the observed frequency of cause codes used ($r = 0.4$). This correlation accounts for 16% of the variance in the observed frequency of cause codes used. The number of inspector positions at FRA is determined in part by the frequency of accidents in each cause code, so perhaps this is a case of positive feedback.

Obviously, these two factors (number of cause codes, number of inspectors) could each underlie the assignment of cause codes to accidents and together could account for a substantial amount of the variance in observed assignment of cause codes. There are a number of ways in which to combine information including various types of averages. The best known averages are the Mean, the Geometric Mean, and the Harmonic Mean (HM). Because it produced the largest correlation between expected frequencies and observed frequencies, the HM was used to combine the two measures. (It should be noted that there was no other rational justification for using the HM.) The relationship between observed and expected frequencies (based on the HM of the proportion of cause codes and the revised proportion of inspectors) is shown in Figure 3.3.

In this case, the correlation is 0.71 ($p < 0.01$) and accounts for more than 50% of the variance in the observed accident frequencies in each cause code.

What are the possible explanations for this surprisingly strong correlation? Several explanations come to mind, but we should first remember that correlation or association does not prove causation. These correlations could be entirely spurious.

The first possible explanation is the well-known range frequency effect in category judgment. Parducci (1974) studied this phenomenon extensively in the 1960s, and the effect is robust. Suffice it to say that category judgments are influenced by the number (range) of categories available and the frequency of those categories. The exact mapping of that effect into this case remains to be worked out, but it is a possible explanation.

It is also possible that there is too little formal process involved in assigning cause codes to accidents, with the result that biases determine a proportion of the assignment of codes. For instance, the railroad employee who makes the assignment of code knows, roughly, how many FRA inspection personnel there are who are concerned

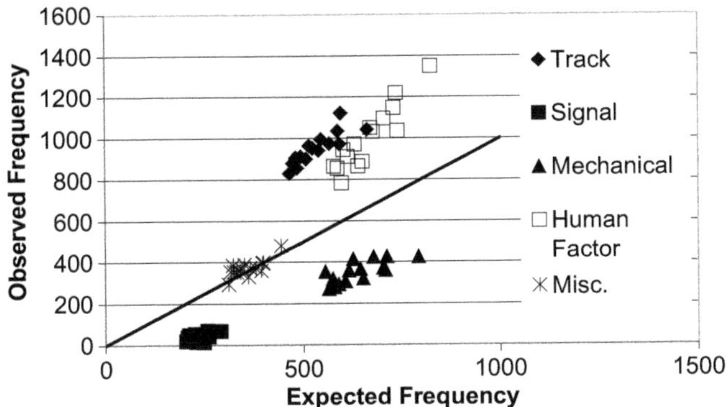

FIGURE 3.3 Observed frequency of accidents in each cause code by expected frequency. Expected frequency is the harmonic mean of available cause codes and number of inspection personnel.

with each type of accident and how many codes there are. In the absence of a formal protocol for determining causation, some codes can be favored or avoided. There may be some bias inherent in code assignment based on who is doing the coding, their background and incentives for assigning particular cause codes.

Of course, like accidents themselves, the assignment of cause codes could have multiple causes. The important point, for our discussion of data issues, is that all data should be questioned and examined carefully to determine possible biases and the validity of the reporting process.

Finally, it should be noted that Grady Cothen, a former senior safety official at FRA, was very disturbed by this discussion of the safety database. He notes that this section "… is great cocktail party fare for the cognoscenti, but it is wrong… Cause codes tend to line up reasonably well, on average, with narratives, and with the few NTSB reports, and the more numerous FRA field investigation reports. Railroads use these data too, and while they have lapses, for sure, they have no incentive, over the long run, to take it casually. Nor, in general, would railroad workers allow it, given that they have the inspectors' numbers on speed dial" (Cothen, 2023, personal communication).

Personally, I do not believe that cause codes are heavily biased by the factors noted above. I believe that the safety database is a valuable tool and that it should not be easily altered. However, there is value in discussing such issues and considering ways to improve data collection.

WHAT IS THE RELATIONSHIP BETWEEN ACCIDENT FREQUENCY AND COST?[2]

Collisions, derailments, fires, explosions, acts of God, or other events involving the operation of railroad on-track equipment (standing or moving) and causing reportable damages greater than the reporting threshold for the year in which the accident/incident occurred must be reported using Form FRA F 6180.54. Table 3.4 shows the reporting thresholds for selected years.

TABLE 3.4
Reporting Thresholds for 1978, 2006–2009, 2017

Year	Threshold
1978	$2,300
2006	$7,700
2007	$8,200
2008	$8,500
2009	$8,900
2017	$10,700

Obviously, what is reported to the FRA depends on the reporting threshold for that year, and this too affects the safety picture presented by the FRA database. This section explores the relationship between accident cost and accident frequency. This is important because many decisions concerning what safety issues need attention, or are under control, are often based on changes in the frequency of particular accidents. Risk is also partly defined by event frequency.

Figure 3.4 shows the frequency of Human Factors accidents as a function of accident cost for the year 2006. Each point on the abscissa is a midpoint for the accident cost since the data were reported as a range of values (e.g., the range $100,000–$199,000 would be represented by the midpoint $150,000).

The point with the highest frequency (791) has a nominal cost of $28,850 because this data bin includes all the accidents above the threshold of $7,700 for 2006 but less than <$50,000. It is clear that the frequency of Human Factors accidents is not a linear function of accident cost. The line represents a power function fit to the data. There is a statistically reliable correlation between the power function and the data $(r = -0.9774, df = 9, p < 0.01)$.

Similar data are presented in Figures 3.5 and 3.6 for collisions and derailments. In both cases, the correlations between the data and power functions are statistically reliable.

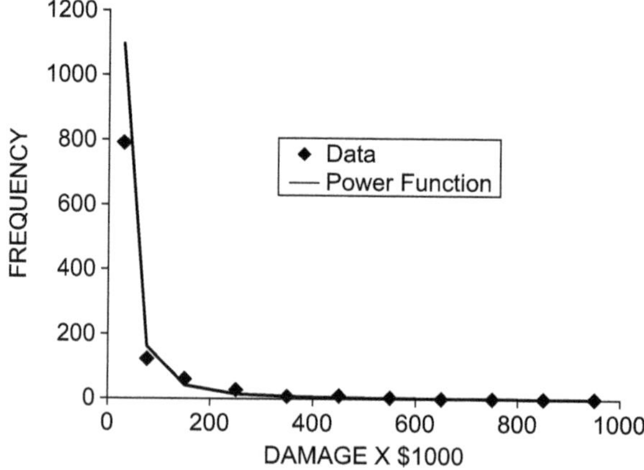

FIGURE 3.4   Frequency of human factors accidents as a function of accident cost.

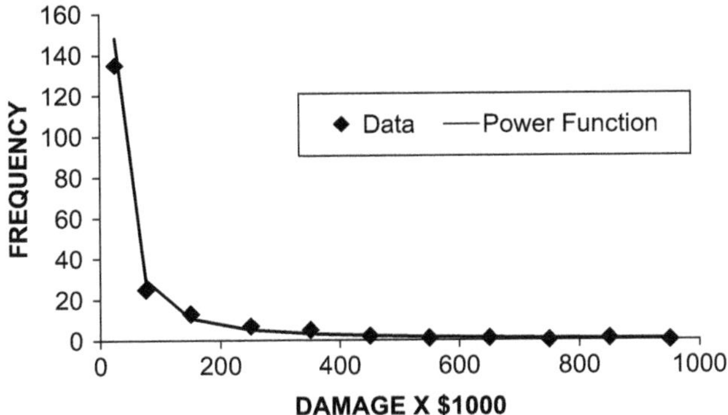

FIGURE 3.5 Frequency of collisions as a function of accident cost.

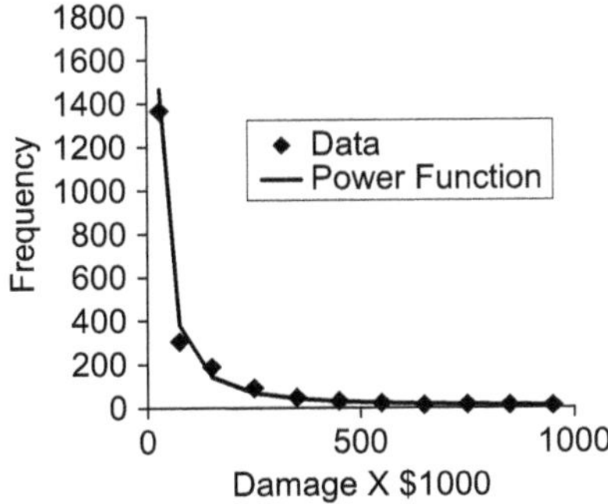

FIGURE 3.6 Frequency of derailments as a function of accident cost.

It is widely accepted that for every fatality in a particular setting there are many more serious injuries, minor reportable accidents, accountable accidents, and unsafe acts (Multer, 2002a, b). This is often illustrated with Heinrich's triangle (1959), shown in Figure 3.7.

This is a principle that shall be encountered again in the discussion of close calls in the railroad industry (see Chapter 9). In the current context, reportable data on accidents and injuries may not be the best indicator of safety risks, and the accident cost data are highly suggestive of this. Decisions about safety risks are often determined by the frequency of events and the damage done by the event. In fact, risk can be defined as the product of event probability and cost (Kumamoto and Henley, 1996). The selection of issues to address is often framed in these terms by examining the frequency or probability of events and their costs in a matrix as shown in Table 3.5.

In Table 3.5 the Harm and Frequency marginal are ranked from least (1) to most (4). Judgment concerning the boundaries for ranks is informed by expert opinion and may vary for different circumstances. This is also the case for the products of the marginals in the matrix. Decisions concerning the allocation of resources to safety

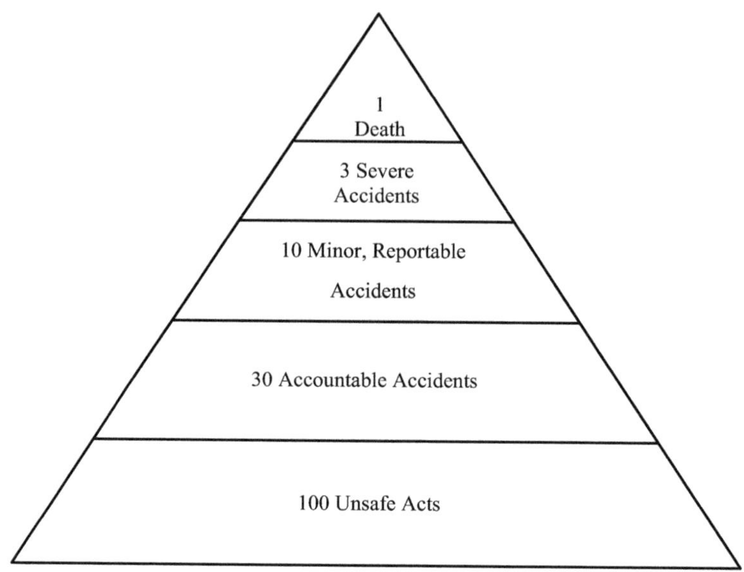

FIGURE 3.7   An example of Heinrich's triangle. The numbers are hypothetical.

TABLE 3.5

A Harm × Frequency (Probability) Matrix

		Harm ($$)			
		1	2	3	4
Frequency	1	*1*	*2*	*3*	*4*
	2	*2*	*4*	*6*	*8*
	3	*3*	*6*	*9*	*12*
	4	*4*	*8*	*12*	*16*

risks can be justified by matrices like Table 3.5. However, in the absence of complete data concerning the frequency and cost of accidents, such decisions may be mistaken.

For instance, if one extrapolates the data from Figures 3.4–3.6 to costs below the reporting threshold (this is, of course, a risky thing to do), one can see the possibility that safety costs are underestimated by the threshold truncated data. The extrapolation is shown in Figure 3.8.

Accidents below the threshold are "Accountable Accidents," meaning that the railroads keep records for these accidents on site, but do not report them to the FRA. Accountable accidents can be audited by the FRA, but this requires that an FRA inspector go to the site where the information is stored to review paper or digital records by hand. This is time-consuming and expensive, so such audits are not done on a systematic basis.

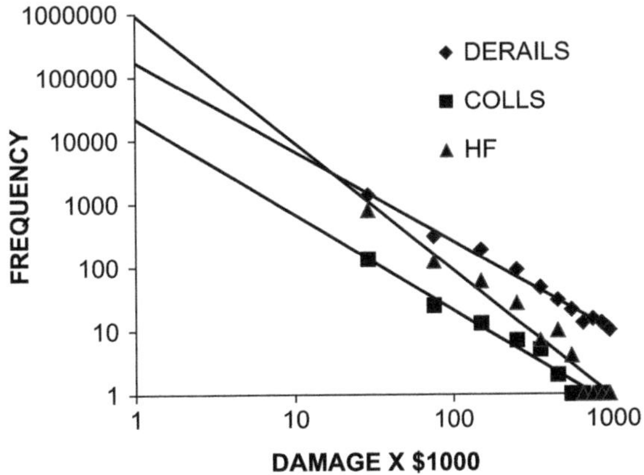

FIGURE 3.8 Frequency of derailments, collisions and human factors accidents as a function of cost. The axes are logarithmic to linearize the power functions of Figures 3.4–3.6.

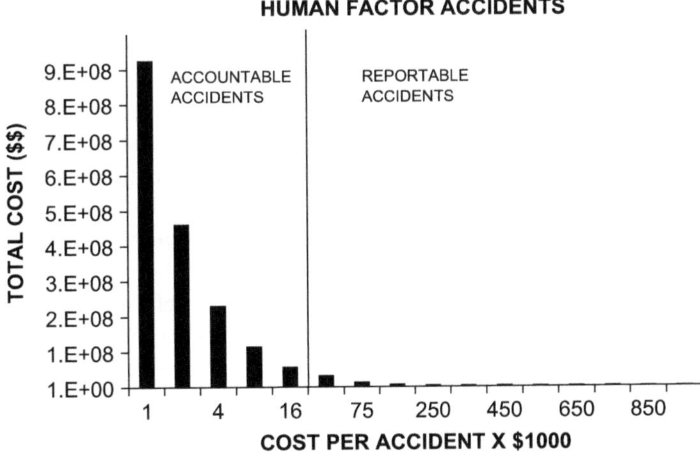

FIGURE 3.9 Total cost ($$) for accountable and reportable human factors accidents. See text for details.

Figure 3.9 shows the putative total cost of accountable accidents and the actual total cost of reportable accidents based on the extrapolation of reportable Human Factor accidents in Figure 3.8. It is clear that, if a power function also applies to the accountable accidents, the greatest reductions in safety risk and cost would be achieved by allocating more resources to accountable accidents.

THE CONCEPT OF RISK

The concept of risk arises and is applied in several ways in this book. There are multiple definitions of risk, and it is important that the reader be aware of which definition is in use and what that definition means in the context of its use.

This book is about railroad safety, and the relationship between safety, risk and harm needs to be clearer. There are many uses of the term "risk", but few people think of risk in the engineering sense of the product of a frequency and a harm (cost) as shown in Table 3.5 above. The concept of risk is useful, but unless it is clear what is meant, it will just lead to confusion. "Safety" is another term that can confuse if it is not stated clearly what is meant, and how it relates to risk. A system is safe, not because there are no risks, but because the risks are understood (assessed) and managed. There are multiple risks associated with any particular outcome (e.g., a derailment that costs <$10,000), and modern thought on accident causation recognizes that all accidents have multiple causes. This notion provides the rationale for discussing risk categories such as derailments in terms of what is done to prevent (reduce the likelihood) and reduce the consequences of risks. However, it now well-recognized that few risks in a system are revealed by accident statistics. Accidents provide a retrospective view of risk, and risk management that only relies on accident statistics is purely reactive. Good risk management requires a proactive approach. There are risks that are revealed by FRA inspection data, accountable accidents (accidents that fall below the reporting dollar threshold), close call reports, etc. Linking risk to accident reports is necessary, but not sufficient, to properly assess and manage risk.

In the social sciences, risk is often conceptualized as a set of probabilities associated with decision outcomes (Krantz et al., 1971, p. 124). In statistics, outcomes in a joint distribution that deviate from the expected values (as determined by the products of marginal values) suggest an association between the variables (e.g., χ^2 test, Hays, 1963), which can be interpreted as risk in the probabilistic sense (i.e., variable Y is more probable as variable X increases or decreases). In epidemiology, risk is a deviation of event rates in a population with a risk factor (exposure to a condition) relative to event rates in a population without exposure to that risk factor (Lilienfeld and Lilienfeld, 1980). The epidemiological and statistical definitions of risk are often called relative risk because the risk is calculated relative to some alternative. In the chapter on fatigue (Chapter 7), the Hursh et al. (2006, 2008) study takes the statistical approach since risk was defined as the ratio of the proportion of accidents at a fatigue level to the proportion of work time at that fatigue level. Generalized to exposure to any risk:

$$\text{Risk ratio} = \frac{\left(\text{Accidents at an exposure magnitue}\right)/\left(\text{Total number of accidents}\right)}{\left(\text{Work time at an exposure magnitude}\right)/\left(\text{Total work time}\right)}. \quad (3.1)$$

In this approach, accident risk due to a magnitude of exposure (e.g., high fatigue) is defined by comparison to duration of exposure (work time at high fatigue). If accident risk is not influenced by exposure magnitude, a plot of risk as a function of exposure magnitude should be a line with a slope of zero. If the slope of that line is statistically greater than zero, accident risk is an increasing function of exposure magnitude. A slope that is less than zero indicates that accident risk is a decreasing function of exposure magnitude.

By contrast, accident risk in epidemiology is defined by the rate of accidents with a magnitude of exposure (e.g. high fatigue) relative to the rate of accidents without that magnitude of exposure (to high fatigue):

$$\text{Risk ratio} = \frac{\text{Accident rate with a magnitude of exposure}}{\text{Accident rate without that magnitude of exposure}}. \tag{3.2}$$

Here, accident rate is the number of accidents with a magnitude of exposure (e.g., high fatigue) divided by the number of hours working without exposure to that magnitude of exposure. A risk ratio of one would indicate that risk is neutral relative to the exposure. Risk ratios greater than one would indicate an increase in risk accruing to the exposure, and ratios less than one would indicate a decrease in risk accruing to the exposure. Statistical methods (Ahlbom, 1993; Fleiss et al., 2003) are used to determine that values of the risk ratio that are different from one are reliable.

Finally, in engineering, risk within a system is defined as a collection of pairs of likelihoods and costs (Kumamoto and Henley, 1996). A common engineering definition of risk is the product of probability and cost. Risk in this view is the expected value which is also the statistical definition of the mean of a distribution (Hays, 1963) and a determinate of response bias in signal detection theory (see Chapter 5). Accident frequencies can be converted to probabilities using the Poisson probability law if the assumptions of that law are met (see Daniel, 1978, p. 72; Gertler et al, 2013, p. 63); also see Chapter 4 Exposure). The probability of one or more accidents $[p(\text{acc} \geq 1)]$ in a time period, T, is given by

$$p(\text{acc} \geq 1) = 1 - e^{-\lambda_T}, \tag{3.3}$$

where λ is the mean number of accidents in T. The probability of an accident given an exposure or magnitude of exposure (conditional probability) can be used in conjunction with harm estimates (conditional accident costs) to calculate risk (see Fatigue Chapter 7).

Risk in a system can be formally defined as a function of a set of ordered pairs of likelihoods and outcomes. This emphasizes the multicausal nature of accidents and how that relates to any safety program such as the R&D program at FRA. This can be functionally stated in a way suggested by Kumamoto and Henley:

$$\text{Risk} \equiv \left\{ (L_1, O_1), \ldots, (L_i, O_i), \ldots, (L_n, O_n) \right\}, \tag{3.4}$$

where O_1 is a train derailment, O_2 is a train collision, O_3 is a trespasser casualty, O_4 is a grade crossing accident, O_5 is a train collision with an object not at a grade crossing, O_6 is all other outcomes, L_1 is the likelihood of a train derailment, L_2 is the likelihood of a train collision, etc.

However, train derailment events have different outcomes. Some may simply involve wheels on the ground of the locomotive. Others may involve minor injuries (O_{11}), loss of life (O_{12}), destruction of property (O_{13}), damage to railroad infrastructure (O_{14}) and/or rolling stock (O_{15}), release of hazardous chemicals (O_{16}), wheels on the ground (O_{17}), or all of the above. Each of these outcomes has a different likelihood profile, hence

$$\text{Risk} \equiv \left\{ \left(L_{ij}, O_{ij} \mid i = 1 \ldots n; \, j = 1 \ldots m \right) \right\}. \tag{3.5}$$

Risk ultimately involves risk management. Just as different types of accidents can each have several different outcomes, the management of each type of accident can be handled in different ways. Train derailments in rail yards are often "wheels on ground". These derailments are often due to human error in which switches are not relined. Alternative solutions can include the discipline of responsible employees, training, better signage, changes in operating practices, etc. Each alternative solution will mitigate some, but not all of the risk, and will have a different associated cost. The alternatives are usually evaluated with some form of cost/benefit analysis which estimates the utility (U) of each. Thus

$$\text{Risk} \equiv \left\{ \left(L_{ij}, O_{ij}U_{ij} \,|\, i = 1 \dots n; \, j = 1 \dots m \right) \right\}. \tag{3.6}$$

The final element for evaluating system risk according to Kumamoto and Henley is causal scenario (CS).

$$\text{Risk} \equiv \left\{ \left(L_{ij}, O_{ij}U_{ij}CS_{ij} \,|\, i = 1 \dots n; \, j = 1 \dots m \right) \right\}. \tag{3.7}$$

The causal scenario specifies the causes of O_i and the sequence of events for O_i. Causes can include issues with track, signals, rolling stock, software, information, the environment, and human factors. This links risk assessment (the likelihood of an occurrence) with the management of risk. A detailed application of this approach to fatigue can be found in Raslear (2010).

NOTES

1 This section is based on an unpublished white paper produced by the author (Raslear, 2008) for analysts in FRA's Office of Safety.
2 This section is based on a presentation to analysts in the FRA's Office of Safety in 2008, see Raslear (2008).

REFERENCES

Ahlbom, A. (1993). *Biostatistics for epidemiologists.* Boca Raton, FL: Lewis Publishers.

Daniel, W. (1978). *Biostatistics: A foundation for analysis in the health sciences.* New York: Wiley.

Federal Railroad Administration (n.d.). *Safety data website.* https://safetydata.fra.dot.gov/OfficeofSafety/publicsite/Query/TenYearAccidentIncidentOverview.aspx

Fleiss, J. L., Levin, B., & Paik, M. C. (2003). *Statistical methods for rates and proportions* (3rd edn). Hoboken, NJ: Wiley.

Gertler, J., DiFiore, A., & Raslear, T. (2013). *Fatigue status of the U.S. Railroad Industry.* (Report No. DOT/FRA/ORD-13/06). Washington, DC: U.S. Department of Transportation.

Gordis, L. (2014). *Epidemiology* (5th edn). Philadelphia, PA: Elsevier.

Hays, W. L. (1963). *Statistics for psychologists.* New York: Holt, Rinehart and Winston.

Heinrich H. W. (1959). *Industrial accident prevention: A scientific approach* (4th edn). New York: McGraw-Hill.

Hursh, S., Raslear, T., Kaye, S., & Fanzone, J. (2006). Validation and calibration of a fatigue assessment tool for railroad work schedules, summary report (Report No. DOT/FRA/ORD 06/21). Washington, DC: U.S. Department of Transportation. https://www.fra.dot.gov/eLib/details/L02535

Hursh, S., Raslear, T., Kaye, S., & Fanzone, J. (2008). Validation and calibration of a fatigue assessment tool for railroad work schedules, final report (Report No. DOT/FRA/ORD 08/04). Washington, DC: U.S. Department of Transportation. https://www.fra.dot.gov/eLib/details/L01597

Krantz, D., Luce, R., Suppes, P., & Tversky, A. (1971). *Foundations of measurement* (Vol. 1). New York: Academic Press.

Kumamoto, H., & Henley, E. J. (1996). *Probabilistic risk assessment and management for engineers and scientists* (2nd edn). New York: The Institute of Electrical and Electronics Engineers.

Lilienfeld, A. M., & Lilienfeld, D. E. (1980). *Foundations of epidemiology* (2nd edn). New York: Oxford University Press.

Multer, J. (2002a). *Improving safety through understanding close call events.* Paper presented at the meeting of the *Federal Railroad Administration Close Call Steering Committee,* Washington, DC.

Multer, J. (2002b). Improving safety through understanding close calls. In J. Saks, J. Multer, & K. Blythe (Eds.), *Proceedings of the human factors workshop: Improving railroad safety through understanding close calls* (Report No. DOT/FRA/ORD-04/03, pp. 113–120). Washington, DC: U.S. Department of Transportation.

Parducci, A. (1974). Contexual effects: A range-frequency analysis. In E. C. Carterette & M. P. Friedman (Eds.), *Handbook of perception: Psychophysical judgment and measurement* (Vol. 2, pp. 127–141). New York: Academic Press.

Raslear, T. G. (2008). *Some observations concerning the data in FRA's accident/incident database.* Unpublished PowerPoint presentation.

Raslear, T. G. (2010). *Fatigue accident risk management.* Unpublished PowerPoint presentation.

Reinach, S., & Gertler, J. (2001). *An examination of railroad yard worker safety* (Report No. DOT/FRA/ORD-01-20). Washington, DC: U. S. Department of Transportation.

United States Government Accountability Office. (2018). *Grade-Crossing Safety: DOT should evaluate whether program provides states flexibility to address ongoing challenges* (Report No. GAO 19-80). Washington, DC: U.S. Government Accountability Office.

4 Exposure

Exposure, in the context of accident or injury risk, refers to the extent to which employees are subjected to conditions which may affect risk. In occupational settings exposure is often expressed as the amount of time employees work in conditions that may affect risk. This is often normalized as the number of accidents or injuries per 200,000 employee-hours (100 employees working 40 h/week for 50 weeks) for comparison across conditions in which different numbers of employees work for different amounts of time. The number of accidents or injuries is a frequency or count variable, which can be statistically characterized by the Poisson probability law. "The Poisson distribution is employed when counts are made of events or entities that are distributed at random in space or time." (Daniel, 1978, p. 72). As noted in Chapter 3, if the assumptions of that law are met, the probability of one or more accidents [$p(\mathrm{acc} \geq 1)$] in an exposure, T, is given by

$$p(\mathrm{acc} \geq 1) = 1 - e^{-\lambda_T}, \tag{4.1}$$

where λ is the mean number of accidents in T. T is usually taken as a time duration but can be a spatial exposure or other metric such as employee-hours. It is characteristic of the Poisson law that doubling T results in a doubling of the number of incidents being counted ("The probability of the single occurrence of the event in a given interval is proportional to the length of the interval" (Daniel, 1978, p. 72)).

BACKGROUND

The genesis of this chapter is in two requests to the author from the railroad industry and from the Federal Railroad Administration's (FRA) Office of Railroad Safety for help in interpreting time of day accident data involving maintenance of way (MOW) workers and signalmen. Previously, the Human Factors Research Division in FRA's Office of Research and Development with the Office of Railroad Safety had funded a project at the Volpe National Transportation Systems Center (Volpe) to examine the use of personal electronic devices in accidents involving maintenance of way (MOW) workers and signalmen (Guthy et al., 2014). As part of that research project, Volpe looked at the circumstances surrounding MOW and signalman accidents and found that human factors causes were over-represented relative to all railroad employees (51% vs. 38%). This, and the two requests for help, caused me to look very carefully at the Fatality Analysis of Maintenance-of-way Employees and Signalmen (FAMES) Committee's[1] time of day accident data since it is documented that, in the railroad industry, fatigue increases the risk of human factors caused accidents (Raslear et al., 2011, 2013), and fatigue is known to vary with time of day (Raslear et al., 2013).

The original FRA report, *"Is There a Time of Day Effect for Maintenance of Way Worker and Signalman Fatalities? Why Risk Exposure is Important"*, was not

DOI: 10.1201/9781003500018-6

published because of objections by the FAMES Committee. Subsequently, I funded a project with Volpe to examine how time of day affects injury risk of railroad maintenance of way employees and signalmen (roadway workers). That project analyzed 15,654 serious roadway worker injuries between 1997 and 2014 and found that nighttime work is more hazardous for roadway workers than daytime work. That project resulted in a report published in the *Journal of Safety Research* (Calabrese et al., 2017).

The purpose of this chapter is to demonstrate the importance of exposure in the analysis of frequency data which is often the basis of accident and injury analyses. Specifically, this chapter analyzes fatality frequencies for maintenance of way (MOW) workers and signalmen in relation to an exposure measure in order to raise awareness of at-risk activity and contribute to railroad safety. We analyzed data included in the FAMES Committee report *Fatal Accident Patterns–Hours of Day* (FAMES, August 1, 2013). The FAMES Committee obtained data on 39 fatal accidents, accounting for 41 fatalities, which occurred between the implementation of the Roadway Worker Protection Rule in 1997 (49 CFR § 214 Part C) and January 1, 2012.

In this chapter, we normalize the fatalities data presented in the FAMES Committee report and present risk ratios at different hours of the day.

THE FAMES COMMITTEE

The FAMES Committee was formed by FRA in collaboration with railroad labor and management representatives to form an ad-hoc committee to review roadway worker fatalities. FAMES is a voluntary, consensus-based committee focused on identifying risks, trends, and factors impacting roadway worker safety. Members include safety representatives from rail labor organizations, railroads, and federal regulators. FAMES periodically issues findings and recommendations based upon its review of available safety data (FAMES Committee, 2012). The FAMES Committee focuses its analysis on two groups of railroad workers: MOW employees and signalmen.

MOW workers build and maintain the tracks, bridges, buildings, and other structures on the railroad. MOW work has two fundamental job classifications, production and non-production. Production jobs involve either track construction or bridge and building construction. Employees with non-production jobs are responsible for inspection, maintenance, and infrastructure repair. Railroads typically assign non-production workers to a specific geographic area, which may encompass several hundred miles end-to-end; they often support the work of production crews when on their territory. Non-production workers also may handle emergency problems at night and on rest days (Gertler and Viale, 2006a).

Signalmen work two fundamental types of jobs: maintenance and construction. Signal maintainers inspect and certify the functioning of the signal and communication equipment on a specific track territory in a defined geographic area. The maintainer is also responsible for making minor repairs found during the inspection. In contrast, signalmen on a construction job can work anywhere on the railroad's system; they likely travel long distances on their own time to reach the construction sites. Most signalmen work on wayside signal equipment, but a limited number work on communications equipment, such as radios and antenna systems (Gertler and Viale, 2006b).

FRA is concerned about the on-the-job safety of MOW workers and signalmen. Both of these roles require individuals to work outside year-round in all types of weather conditions and over varying terrain. Individuals in these roles work on a regular daytime schedule to cover non-yard territory, on a shift work schedule to cover the yard around the clock and are responsible for responding to emergencies at night and on weekends. MOW workers and signalmen may be exposed to an extremely hazardous environment because of the nature of the work and the operation of trains and other on-track equipment within the work environment (Gertler and Viale, 2006a, b). The introduction of Roadway Worker Protection regulations in 1997 has helped to reduce the risk to these groups of railroad workers of being struck by trains and other on-track equipment. Nevertheless, because of the hazardous work environment of MOW workers and signalmen, safety is always a critical issue (Gertler and Viale, 2006a, b).

METHODS

The objective of this chapter is to demonstrate that fatality frequencies for MOW workers and signalmen should be analyzed in relation to an exposure measure in order to determine whether the risk of a fatality varies as a function of time of day (TOD). The FAMES Committee report analyzed the raw frequencies of fatal accidents at different TODs. We analyzed these frequencies in relation to a measure of exposure. Specifically, we normalized the accident frequency data presented in the FAMES Committee report by using data on the proportion of employee-hours worked at different TODs. Then, we calculated the relative risk of a fatality and drew conclusions about the relative risk of MOW workers and signalmen at different TODs.

We normalized the fatalities data in order to compare the time of day on equivalent bases. Normalizing the data takes into account sources of variation (such as the differences in numbers of individuals working at each hour of the day) that may influence the results; it assists in interpreting differences between the fatality frequencies at different times of the day.

As noted above, we analyzed fatal accidents. The variable *fatal accidents* is a count variable. Count variables reflect the number of occurrences of an event in a specific amount of time. Count variables can only take on positive integer values or zero, because an event cannot happen a fractional or negative number of times.

The Poisson distribution is one way to model count variables that occur randomly in a given interval of time. The first question we asked was whether the fatal accident data could be adequately modeled by the Poisson distribution. In the Poisson process, the "...probability of the single occurrence of the event in a given interval is proportional to the length of the interval..." (Daniel, 1978, p. 72). The fatality data is organized in one-hour bins, which means that the probability of a fatality at each TOD should be equal, provided that the fatal accident data are adequately modeled by the Poisson distribution and the same number of employee-hours (exposure) are worked at each TOD.

We obtained baseline data about the number of MOW workers and signalmen working at each hour of the day from two prior FRA studies. These studies examined the work schedules and sleep patterns of MOW workers and signalmen. Participants used a daily log to record sleep and work periods on both regular workdays and planned days off for a 2-week period (Gertler and Viale, 2006a, b).[2]

Participants' logs (i.e., work-rest diary databases) contained data about the days of the week that they worked in a 2-week period, and the start and end times of their shifts. We used this information to calculate the number of individuals working at each hour of the day on each day of the week (e.g., the number of MOW workers and signalmen working the first Monday of the diary period at 0700). Then, we determined the number of employee-hours worked at each hour of the day, irrespective of the day of the week (e.g., the number of MOW workers and signalmen, combined, working at 0700).

From the work-rest diary databases, we obtained usable data from 639 individuals (254 MOW workers and 385 signalmen).

DATA ANALYSIS

We used the Poisson distribution to model the MOW worker and signalmen fatalities data. We tested the goodness-of-fit of the data to the Poisson distribution with a χ^2 test and obtained 95% confidence intervals (CIs) for each of the 24 data points in the fatalities data.

We computed a risk ratio using the fatalities frequency data and data from the work-rest diary databases. The risk ratio reports the likelihood that an event will occur, given a certain level of exposure to the risk factor. In the case of the fatalities data, we were interested in the likelihood of a fatality occurring, given the number of MOW and signalmen employee-hours worked at a particular time of day.

In our analysis we first convert the fatalities frequency data into the proportion of fatalities as a function of TOD. We then present the data on the proportion of employee-hours worked for the same TODs. A risk ratio is calculated by dividing the proportion of fatalities by the proportion of employee-hours for each TOD. Because there are 41 fatalities distributed among 24 h, there are many hours of the day with no fatalities. To reduce the effect of random variations in the data (i.e., noise), the data for proportion of fatalities and proportion of employee-hours was re-binned into 3-h increments to eliminate zeros in the proportions. A risk ratio for each TOD was then re-calculated from the re-binned data.

RESULTS

ARE FATAL ACCIDENTS ADEQUATELY MODELED BY THE POISSON DISTRIBUTION?

The data on frequency of fatalities as a function of TOD is shown in Figure 4.1[3] (see Table A.3).

We used the Poisson distribution to model the MOW worker and signalmen fatalities data. In the Poisson distribution, the probability of an event x, $p(x)$, is

$$p(x) = \frac{e^{-\lambda}\lambda^x}{x!}, \tag{4.2}$$

where λ is mean number of fatalities in an hour.

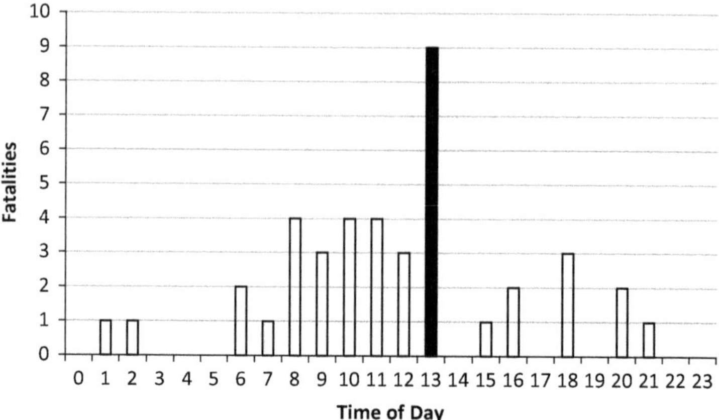

FIGURE 4.1   Fatalities as a function of time of day. After (FAMES Committee, 2012).

TABLE 4.1

Number of Hours of Day and Relative Frequencies Associated with Fatality Frequencies

Fatality Frequency	Number of Hours with Given Frequency	Relative Frequency
0	9	0.375
1	5	0.208333
2	3	0.125
3	3	0.125
4	3	0.125
5	0	0
6	0	0
7	0	0
8	0	0
9	1	0.041667
10	0	0
Mean	1.708333	

To obtain the Poisson distribution of the data, we re-organized the data to show the number of hours of the day that were associated with different frequencies of fatalities. Then, we calculated relative frequencies of fatalities and the mean number of fatalities.

As shown in Table 4.1, there were 9 h of the day that had zero fatalities, 5 h of the day that had one fatality, 3 h of the day that had two fatalities, etc. The mean number of fatalities for any given hour of the day was 1.71. We then fitted a Poisson distribution with this mean value ($\lambda = 1.71$) by calculating the expected frequencies of the distribution from Equation (4.1) (see Table 4.2).

Figure 4.2 depicts these data.

TABLE 4.2

Observed and Expected Number of Hours Associated with Different Fatality Frequencies

Fatality Frequency	Expected Number of Hours (Poisson)	Observed Number of Hours
0	7.427867	9
1	9.148337	5
2	7.814204	3
3	4.449755	3
4	1.900416	3
5	0.649309	0
6	0.184873	0
7	0.045118	0
8	0.009635	0
9	0.001829	1
10	0.000312	0

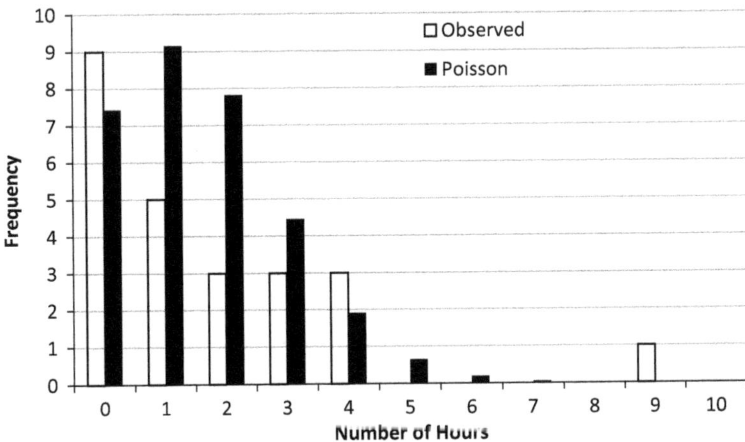

FIGURE 4.2   Observed and expected number of hours associated with different fatality frequencies.

Next, we wanted to see if the data were Poisson distributed. The chi-squared (χ^2) test is the usual statistical test for goodness-of-fit (Hays, 1963). A requirement of the χ^2 test is that the expected frequency (labeled Poisson in Figure 4.2) in each bin is at least five. To meet this requirement, we truncated the distributions into four bins as shown in Table 4.3.

Figure 4.3 depicts these data.

To determine whether the Poisson distribution was a good fit to the data, we conducted a χ^2 test for goodness-of-fit:

TABLE 4.3
Observed and Expected Number of Hours
Associated with Four Bins of Fatality Frequencies

Fatality Frequency	Expected Number of Hours (Poisson)	Observed Number of Hours
0	7.427867	9
1	9.148337	5
2	7.814204	3
3+	7.241246	7

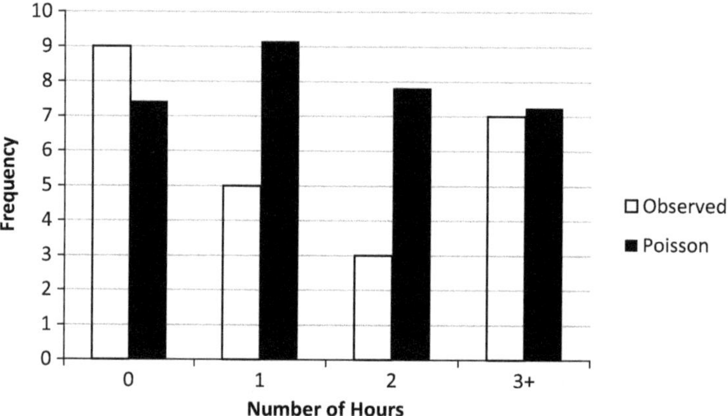

FIGURE 4.3 Observed and expected number of hours associated with four bins of fatality frequencies.

$$\chi^2 = \Sigma \frac{(\text{observed} - \text{expected})^2}{\text{expected}}. \tag{4.3}$$

The χ^2 value is small when the observed and expected frequencies are close and large when they are not close; thus, large χ^2 values indicate that the data do not follow a Poisson distribution.

To determine whether the χ^2 is significant, we compared the obtained χ^2 of 5.187811 with the critical value. The critical value for two degrees of freedom[4] was 5.99. Our χ^2 did not exceed the critical value for two degrees of freedom. Therefore, the Poisson distribution is a good fit of the data.

Since the Poisson distribution adequately models the fatalities frequency data, we established 95% confidence intervals for the Figure 4.1 data (Ahlbom, 1993, Appendix 1). This is shown in Figure 4.4.

With the exception of 1300h (9 fatalities) versus 0000, 0300, 0400, 0500, 1400, 1700, 1900, 2200 and 2300h (0 fatalities), all of the confidence intervals overlap, indicating that further analysis of this data set is warranted. As indicated above, the

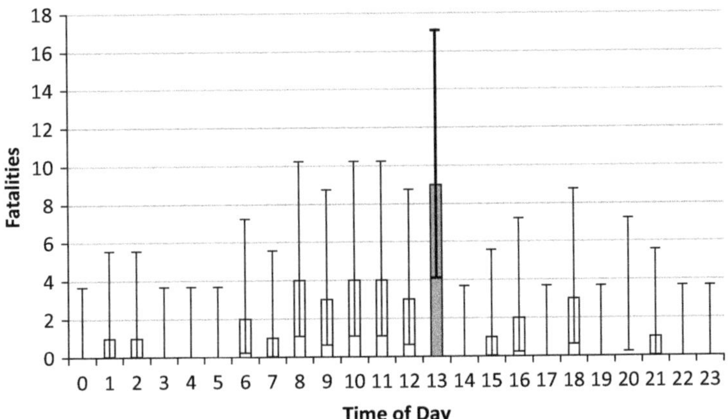

FIGURE 4.4 Fatalities as a function of time of day and 95% confidence intervals.

Poisson analysis assumes that exposure in all time intervals is equal. The observed inequality could be due to any number of other exposure variables, including the number of employee-hours worked at different TODs. This is explored in the next section.

EMPLOYEE-HOURS AS AN EXPOSURE MEASURE

Figure 4.5 shows the same data as in Figure 4.1 converted to proportions (see Table B.1). This was done as a first step to calculating the risk ratios described below.

Figure 4.6 shows the proportion of employee-hours as a function of TOD.[5] Figure 4.6 is based on the work/rest diary studies for MOW workers (Gertler and Viale, 2006a) and signalmen (Gertler and Viale, 2006b). These diary studies are representative samples of MOW worker and signalman populations. The proportion of employee-hours in Figure 4.6 are statistical estimates based on those samples. We are using these statistical estimates of the proportion of employee-hours instead of employee-hours worked because data for employee-hours worked during the period of interest (1997–2012) are not available. Note that the proportion of employee-hours is relatively flat from 0700 to 1500h. From 2000 to 0500h, there are very few employee-hours worked.

Using the data in Figures 4.5 and 4.6 we calculated the fatalities risk ratio for each time of day. The fatalities risk ratio is defined as follows:

$$\frac{\text{Proportion of fatalities in time of day bin } x}{\text{Proportion of employee} - \text{hours in time of day bin } x}. \tag{4.4}$$

A risk ratio of 1 means a fatality is no more likely during a particular hour of the day compared to the others. A risk ratio <1 means a fatality is less likely to occur during a particular hour of the day compared to other hours. A risk ratio >1 means a fatality is more likely to occur during a particular hour of the day compared to other hours.

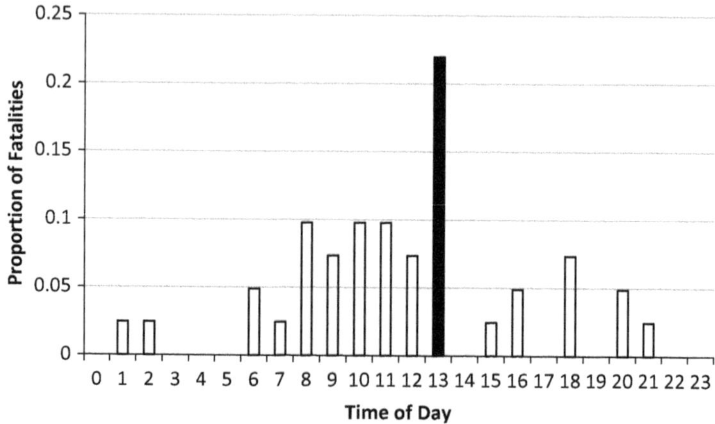

FIGURE 4.5   Proportion of fatalities as a function of time of day.

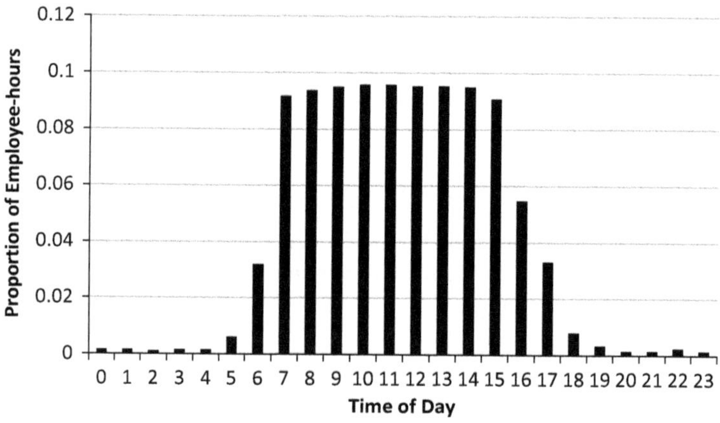

FIGURE 4.6   Proportion of employee-hours as a function of time of day.

Figure 4.7 shows the risk ratio as a function of TOD.

Two points are worthy of note. First, there is an extremely elevated risk of a fatality at 0100, 0200, 1800, 2000 and 2100h. Second, there are TODs for which the relative risk is zero. A meaningful risk ratio cannot be calculated for a particular hour of the day if there were no fatalities during that hour.[6] As shown in Figure 4.7, 9 hours of the day had no fatalities: 0000, 0300, 0400, 0500, 1400, 1700, 1900, 2200, and 2300.

Since there are a considerable number of zeros in this dataset, it is desirable to bin the data to eliminate the zeros. Consequently, we grouped the fatalities data into six bins of 3 h each. The hours of the day associated with each bin are:

0000–0300 (i.e., 0–3),
0400–0700 (i.e., 4–7),
0800–1100 (i.e., 8–11),
1200–1500 (i.e., 12–15),
1600–1900 (i.e., 16–19), and
2000–2300 (i.e., 20–23).

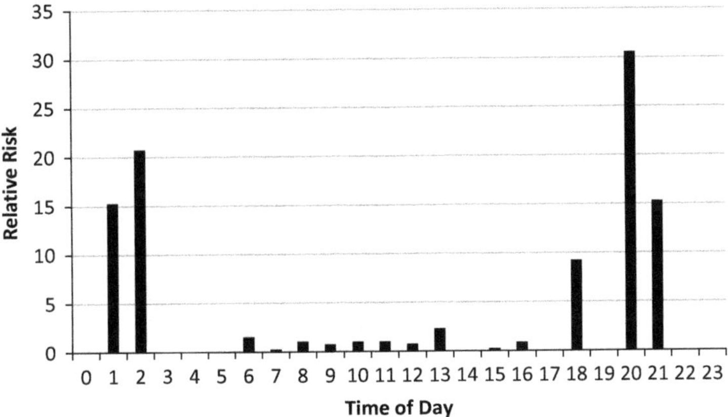

FIGURE 4.7   Risk ratio as a function of time of day.

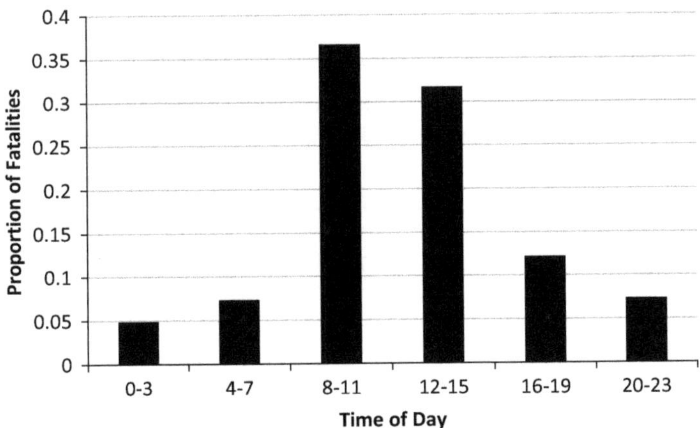

FIGURE 4.8   Proportion of fatalities as a function of time of day.

We re-expressed the frequency data as a proportion of the total number of fatalities in each TOD bin (see Figure 4.8).

The data in Figure 4.8, however, are not easy to interpret because we do not know how many employees worked in each time of day bin. If there were many more employees working between 0800 and 1500 than between 1600 and 0700, then one would expect more fatalities because there was more exposure (i.e., employee-hours) between 0800 and 1600. Although we do not know the number of employee-hours for the data in Figure 4.8, we can calculate a risk ratio (i.e., relative risk) that takes into account the proportion of employee-hours worked in each time bin.

Figure 4.9 shows the distribution of the proportion of MOW workers and signal-man hours as a function of time of day. Note that the bins in Figure 4.9 are the same as in Figure 4.8.

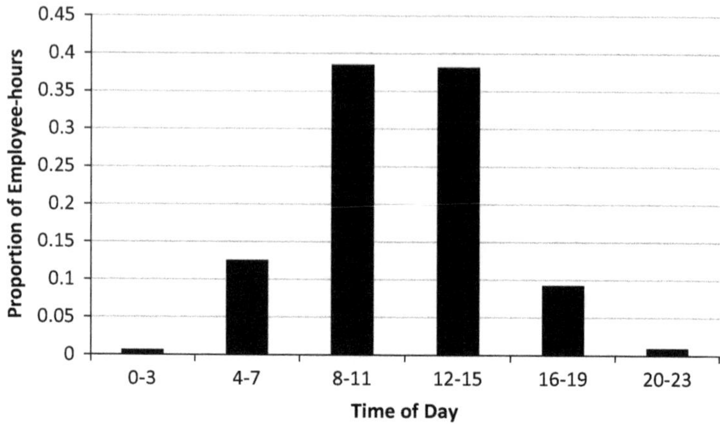

FIGURE 4.9   Proportion of MOW and signalman hours worked as a function of time of day.

TABLE 4.4

Fatality Frequencies and Risk Ratios by Time of Day

Time of Day	Frequency of Fatalities	Proportion of Fatalities	Proportion of Employee-Hours	Risk Ratio
0–3	2	0.048780	0.006694019	7.287175
4–7	3	0.073171	0.125041388	0.585172
8–11	15	0.365854	0.384783704	0.950803
12–15	13	0.317073	0.381285539	0.831590
16–19	5	0.121951	0.092435039	1.319318
20–23	3	0.073171	0.009760311	7.496762
Total	41	1.000000	1.000000000	

A small proportion of MOW workers and signalmen work between the hours of 2000 and 0300. The proportion increases to a peak of almost 0.40 between 0800 and 1500 before declining again after 1500. In other words, the hours of 0800–1500 are associated with the greatest proportions of MOW workers and signalmen performing their work.

Table 4.4 shows the frequency of fatalities, the proportion of fatalities, the proportion of employee-hours and the risk ratio as a function of time of day.

Figure 4.10 shows a plot of the risk associated with each time of day bin. Error bars indicate the 95% confidence intervals around each data point.

It is clear that there is reliably higher risk of a fatality from 2000 to 0300. Risk from 0400 to 1900 is not different from 1, indicating that there is not elevated risk of a fatality during these hours. The slight overlap between time periods 1600–1900 and 2000–2300 indicates that risk may be increasing in the 1600–1900 time period.

Note that with normalization by employee-hour exposure, a totally different picture has emerged. Fatalities are not more likely in the time period following a meal (1300 h). Rather, consistent with circadian rhythms, late night and early morning hours are 7 times more likely to have fatal accidents occur.

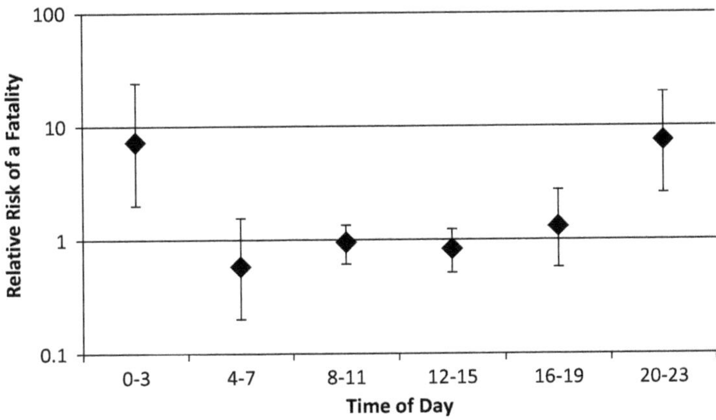

FIGURE 4.10   Relative risk of fatality as a function of time of day.

DISCUSSION

MOW workers and signalmen are known to have fixed daytime work schedules (Gertler and Viale, 2006a, b). Work outside of the hours of 0600 and 1800 is unusual and likely to involve emergency calls. Workers who are called back to work for emergencies, or who are required to continue work because of an emergency, would not have sufficient restorative rest to be fully alert and cognitively effective. Circadian rhythms would further impair alertness and cognitive function. Reaction times slow, lapses in attention increase, and cognitive throughput is reduced under these conditions resulting in an increase in human error (Hursh et al., 2004). Human error, in turn, can result in an accident or fatality (Reason, 1990, 1997).

We analyzed the fatality frequency data using relative risk and proportion of employee hours because we did not have a measure of the number of hours MOW workers and signalmen worked from 1997 to 2012 (i.e., in the time period represented by the fatalities data).

We used the 2006 work-rest diary databases to obtain employee-hours because we knew the accuracy and precision of these data. The employee-hours data were associated with random samples of the U.S. MOW workers and signalmen populations and were statistically valid and representative of those populations at the time of the data collection.

LIMITATIONS

A limitation of using the 2006 employee-hours data as a measure of the number of hours worked by MOW workers and signalmen is that the proportion includes data from only 2006—it might not be representative of the whole time period associated with the fatalities data. Some unknown amount of error could have been introduced into the analyses if 2006 was not representative of the whole time period 1997–2012.

Another limitation of our methodology is that we calculated relative risk, not absolute risk. We were able to obtain a comparison between different risk levels (e.g., a fatality is seven times more likely between the hours of 2000–0300 as compared to

0800–1100), but we are not able to provide a measure of the fatality rate. The fatality rate is a measure of absolute risk and it allows one to calculate the probability of a fatality during a particular time of day.

Additional data are needed to determine the factors that contribute to the increased likelihood of fatal accidents during the late night and early morning hours. It seems likely that a circadian effect may contribute to the increase in relative risk at these times of day, but additional data are needed. The relative risk measures we calculated were associated with large 95% confidence intervals, and we were analyzing a small dataset of fatal accidents. As a result, substantial statistical uncertainty limits the conclusions that can be drawn from these data.

NEXT STEPS

As a next step, non-fatal accident data as a function of time of day should be reviewed. Since there are many more non-fatal accidents, this will improve the statistical reliability of a time of day analysis and help to confirm the results[7] reported here.

Additionally, the causes of the five fatal accidents that occurred during the late night and early morning hours should be examined. A circadian pattern to the fatalities suggests that human error was involved. If human factor causes are associated with these fatalities, the plausibility of a circadian effect is strengthened. This is the same logic that was used to validate the Fatigue Avoidance Scheduling Tool, FAST (Hursh et al. 2008). Non-biological (i.e., mechanical) systems such as track, rolling stock and signals should not exhibit a circadian rhythm.

CONCLUSION

When exposure information about the number of MOW workers and signalmen working at each hour of the day is taken into account, we find that the relative risk of a fatality is significantly increased in the time period from 2000 to 0300. During these hours of the day, the risk of fatality is seven times the risk of fatality at other times of the day. Risk is relatively neutral between the hours of 0400 and 1500 and increases slightly during the 1600–1900 time period.

On the basis of the frequency data alone, one might conclude that 1300 is a particularly dangerous TOD, whereas the normalized data indicate the opposite. This illustrates that it is important to consider events in relation to their relative risk of occurrence on the basis of an appropriate measure of exposure to that risk so as to not reach a false conclusion.

The FAMES Committee report *Fatal Accident Patterns—Hours of Day* (2013) contained the following recommendations:

Roadway Workers should focus on the safety risk of transitioning from a sedentary period to an active work environment.

Railroads should engage Roadway Workers in discussions concerning the spike of fatalities from 1:00 p.m. to 2:00 p.m. to raise awareness and gain a better understanding of the underlying causes.

Roadway Workers should participate in a Job Safety Re-Briefing after extended periods of inactivity, such as meal periods, travel time, etc.

Rather than cautioning MOW workers and signalmen about safety after the noon mealtime, the FAMES Committee should have been emphasizing the dangers of work after the normal daytime shift.

The very low proportion of employee-hours between the hours of 2000 and 0300 translate the few fatalities at that TOD into a high relative risk.

NOTES

1   The FAMES Committee consists of government, railroad labor and railroad management officials who monitor fatalities in railroad MOW workers and signalmen (FAMES Committee, 2012).
2   The two Microsoft Excel datasets associated with these studies are: Railroad Maintenance of Way Employee Background Survey (http://www.fra.dot.gov/eLib/details/L04108) and Railroad Signalmen Background Survey (http://www.fra.dot.gov/eLib/details/L04111).
3   Appendix A contains the raw data associated with Figure 4.1.
4   Usual degrees of freedom for a goodness-of-fit test is $N - 1 = 4 - 1 = 3$. Since we estimated λ from the data, an additional degree of freedom was subtracted from the number of categories (N).
5   Appendix B contains the raw data associated with Figure 4.5.
6   For hours of the day with zero fatalities, the risk ratio is zero. A risk ratio of zero has no meaning.
7   This was accomplished by the Volpe Center and published by Calabrese et al. (2017).

REFERENCES

Ahlbom, A. (1993). *Biostatistics for epidemiologists*. Boca Raton: Lewis.

Calabresse, C., Mejia, B., McInnis, C. A., France, M., Nadler, E., & Raslear, T. (2017). Time of day effects on railroad worker injury risk. *Journal of Safety Research, 61*, 53–64.

Daniel, W. (1978). *Biostatistics: A foundation for analysis in the health sciences*. New York: Wiley.

FAMES Committee. (May 21, 2012). *Introduction to the FAMES committee*. Washington, DC: Federal Railroad Administration. https://www.fra.dot.gov/eLib/Details/L01182.

FAMES Committee. (August 1, 2013). *Fatal accident patterns—hours of day*. Washington, DC: Federal Railroad Administration. https://www.fra.dot.gov/eLib/details/L04711.

Federal Railroad Administration. (2008a). *Railroad maintenance of way employee background survey* [Data file]. https://www.fra.dot.gov/eLib/details/L04108.

Federal Railroad Administration. (2008b). *Railroad signalmen background survey* [Data file]. https://www.fra.dot.gov/eLib/details/L04111.

Gertler, J., & Viale, A. (2006a). *Work schedules and sleep patterns of railroad maintenance of way workers* (DOT/FRA/ORD-06/25). Washington, DC: Federal Railroad Administration. https://www.fra.dot.gov/eLib/details/L01638.

Gertler, J., & Viale, A. (2006b). *Work schedules and sleep patterns of railroad signalmen.* (DOT/FRA/ORD-06/19). Washington, DC: Federal Railroad Administration. https://www.fra.dot.gov/eLib/details/L02537.

Guthy, C., Rosenhand, H., Bisch, A., & Nadler, E. (2014). *Safety of railroad employees' use of personal electronic devices* (DOT/FRA/ORD-14/16). Washington, DC: Federal Railroad Administration. https://www.fra.dot.gov/eLib/details/L05272#p1_z5_gD_lRT_kpersonal%20electronic%20devices.

Hays, W. (1963). *Statistics for psychologists*. New York, Holt, Rinehart and Winston.

Hursh, S., Raslear, T., Kaye, S., & Fanzone, J. (2008). *Validation and calibration of a fatigue assessment tool for railroad work schedules, final report* (DOT/FRA/ORD-08/04). Washington, DC: Federal Railroad Administration. https://www.fra.dot.gov/eLib/details/L01597.

Hursh, S., Redmond, D., Johnson, M., Thorne, D., Belenky, G., Balkin, T., Storm, W., Miller, J., & Eddy, D. (2004). Fatigue models for applied research in warfighting. *Aviation, Space, and Environmental Medicine, 75*, A44–53.

Raslear, T. G., Gertler, J., & DiFiore, A. (2013). Work schedules, sleep, fatigue, and accidents in the US railroad industry. *Fatigue: Biomedicine, Health & Behavior, 1*, 99–115

Raslear, T. G., Hursh, S. R., & Van Dongen, H. P. A. (2011). Predicting cognitive impairment and accident risk. In H. P. A. Van Dongen & G. A. Kerkhof (Eds.), *Progress in brain research: Human sleep and cognition. Part II: Clinical and applied research* (Vol. 190, pp. 155–167). Amsterdam: Elsevier.

Reason, J. (1990). *Human error.* Cambridge, UK: Cambridge University Press.

Reason, J. (1997). *Managing the risks of organizational accidents.* Burlington, VT: Ashgate.

APPENDIX A. MOW AND SIGNALMEN FATALITIES – RAW DATA

Tables A.1 and A.2 list the number of MOW employees and signalmen working during each day of the week and each hour of the day.

TABLE A.1
Week 1   Raw Data from FRA Work-Rest Diaries

Time of Day	Sun	Mon	Tues	Wed	Thurs	Fri	Sat
0:00	1	3	3	3	4	3	2
1:00	1	3	3	3	4	3	2
2:00	1	3	2	2	3	2	1
3:00	1	4	3	3	4	3	1
4:00	1	4	3	3	4	3	1
5:00	4	16	15	15	15	9	4
6:00	16	80	81	81	78	53	20
7:00	28	225	231	231	227	158	32
8:00	29	229	236	236	232	163	34
9:00	30	231	238	239	235	165	35
10:00	31	233	240	241	237	166	35
11:00	31	233	240	241	237	166	35
12:00	31	232	239	240	236	165	35
13:00	31	232	239	240	236	165	35
14:00	31	231	238	239	235	164	35
15:00	31	221	228	229	225	154	35
16:00	27	136	142	143	140	74	33
17:00	22	84	90	91	88	37	28
18:00	12	17	19	20	20	19	17
19:00	3	6	7	8	8	8	7
20:00	0	3	4	4	4	3	1
21:00	0	3	4	4	4	3	1
22:00	1	5	5	5	6	5	2
23:00	1	3	3	3	4	3	2
Sum	364	2,437	2,513	2,524	2,486	1,694	433

TABLE A.2
Week 2　Raw Data from FRA Work-Rest Diaries

Time of Day	Sun	Mon	Tues	Wed	Thurs	Fri	Sat
0:00	1	3	3	3	4	3	2
1:00	1	3	3	3	4	3	2
2:00	1	3	2	2	3	2	1
3:00	1	4	3	3	4	3	1
4:00	1	4	3	3	4	3	1
5:00	5	15	14	14	14	6	1
6:00	20	73	73	72	70	40	2
7:00	36	215	219	218	215	140	7
8:00	37	218	223	222	219	144	8
9:00	38	220	225	225	222	146	9
10:00	39	222	227	227	224	147	9
11:00	39	222	227	227	224	147	9
12:00	39	221	226	226	223	146	9
13:00	39	221	226	226	223	146	9
14:00	39	220	225	225	222	145	9
15:00	39	210	215	215	212	135	9
16:00	35	125	129	129	127	55	7
17:00	28	71	76	76	73	19	4
18:00	16	9	10	10	10	7	3
19:00	5	4	5	6	6	5	3
20:00	0	3	4	4	4	3	1
21:00	0	3	4	4	4	3	1
22:00	1	5	5	5	6	5	2
23:00	1	3	3	3	4	3	2
Sum	461	2,297	2,350	2,348	2,321	1,456	111

TABLE A.3
Descriptive Statistics for Combined Week 1 and
Week 2 Raw Data from Tables A.1 and A.2

Time of Day	Sum	Minimum	Maximum	Mean
0:00	38	0	4	2.53
1:00	38	0.041667	4	2.54
2:00	28	0.083333	3	1.87
3:00	38	0.125	4	2.54
4:00	38	0.166667	4	2.54
5:00	147	0.208333	16	9.81
6:00	759	0.25	81	50.62
7:00	2,182	0.291667	231	145.49
8:00	2,230	0.333333	236	148.69
9:00	2,258	0.375	239	150.56

(Continued)

TABLE A.3 (*Continued*)
**Descriptive Statistics for Combined Week 1 and
Week 2 Raw Data**

Time of Day	Sum	Minimum	Maximum	Mean
10:00	2,278	0.416667	241	151.89
11:00	2,278	0.458333	241	151.90
12:00	2,269	0.5	240	151.23
13:00	2,269	0.541667	240	151.24
14:00	2,259	0.583333	239	150.57
15:00	2,159	0.625	229	143.91
16:00	1,303	0.666667	143	86.84
17:00	788	0.708333	91	52.51
18:00	190	0.75	20	12.65
19:00	82	0.791667	8	5.45
20:00	39	0	4	2.59
21:00	39	0	4	2.59
22:00	59	0.916667	6	3.93
23:00	39	0.958333	4	2.60
Sum	23,807			

APPENDIX B. FATAL ACCIDENTS – RAW DATA

TABLE B.1
**Number and Proportion of Fatalities as a Function of
Time of Day**

Time of Day	Fatality Frequency	Relative Frequency (i.e., Proportion of Fatalities as a Function of TOD)
0:00	0	0
1:00	1	0.02439
2:00	1	0.02439
3:00	0	0
4:00	0	0
5:00	0	0
6:00	2	0.04878
7:00	1	0.02439
8:00	4	0.097561
9:00	3	0.073171
10:00	4	0.097561
11:00	4	0.097561
12:00	3	0.073171

(Continued)

TABLE B.1 (*Continued*)
Number and Proportion of Fatalities as a Function of Time of Day

Time of Day	Fatality Frequency	Relative Frequency (i.e., Proportion of Fatalities as a Function of TOD)
13:00	9	0.219512
14:00	0	0
15:00	1	0.02439
16:00	2	0.04878
17:00	0	0
18:00	3	0.073171
19:00	0	0
20:00	2	0.04878
21:00	1	0.02439
22:00	0	0
23:00	0	0
Sum	41	
Mean	1.708333	
Standard Deviation	2.115762	

5 Signal Detection Theory

I have used Signal Detection Theory (SDT) to analyze data in a number of projects at FRA, including motorist behavior at grade crossings (see Chapter 6), track inspection (see Chapter 10), train conspicuity (see Chapter 6), train horn audibility (see Chapter 6), and wayside track signals. Rather than repeat basic information in the discussion of each of these projects, this chapter will provide a general overview of SDT that can be referenced when the specific topics are described. References at the end of this chapter can provide more advanced and complete descriptions of the theory.

DETECTION OF A SIGNAL IN A BACKGROUND OF NOISE

SDT is a basic model of human information processing that incorporates two major ideas: human sensory systems have the ability to detect stimuli in the environment (sensitivity), and that ability to detect is independent of the decision that a stimulus has been detected (bias). SDT was first formulated to aid in the analysis of radar operator behavior. In this situation, a signal (a spot of light on a radar screen representing an aircraft) must be detected in a background of noise (other spots of light that are due to clouds and other atmospheric conditions), and a decision is made as to whether the spot of light is an aircraft or noise. Signals vary in magnitude as does noise, which makes the detection and decision-making process difficult. In SDT signals and noise are typically represented as overlapping Gaussian (normal) distributions. The extent to which the distributions overlap determines the sensitivity of the observer. Decision-making involves the observer setting a criterion for whether an ambiguous stimulus is considered a signal or noise.

Figure 5.1 shows the basic situation. There are two states of the world: noise and signal, represented by the overlapping Gaussian probability distributions. The distance between the means of the distributions (the peaks of the distributions) indicates the observer's sensitivity or ability to detect a signal. There are also two possible decisions that the observer can make: the stimulus is a signal, or the stimulus is noise. The vertical line in Figure 5.1 is a criterion line. Stimuli which fall on the right side of the criterion are considered to be signals, while stimuli which fall on the left side of the criterion are considered to be noise. The criterion line can be moved to the left or right depending upon the observer's motivation, expectations and attention. Changes in the setting of the criterion do not affect the location of the noise and signal distributions, hence bias (the location of the criterion) is independent of sensitivity. Table 5.1 shows the relationship between observer decisions and the state of the world.

In SDT terminology, if a signal is present and the observer decides that the stimulus is a signal ("Yes" response), a Hit (the correct detection of a signal) has occurred. If a signal is not present and the observer decides that the stimulus is a signal, a False Alarm (the false detection of a signal) has occurred. If a signal is present and the

DOI: 10.1201/9781003500018-7

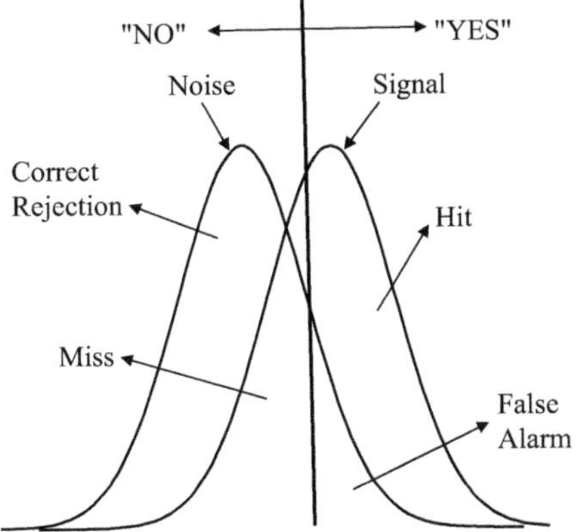

FIGURE 5.1 Observer's responses and signal and noise probability distributions.

TABLE 5.1

Four Possible Outcomes in an SDT Matrix

		Observer Response	
		"Yes"	**"No"**
States of the world	Signal	*Hit*	*Miss*
	Noise	*False Alarm*	*Correct Rejection*

---

observer decides that the stimulus is noise ("No" response), a Miss (the failure to detect a signal) has occurred. If a signal is not present and the observer decides that the stimulus is noise, a Correct Rejection has occurred. Since Hits and Misses are probabilities and come from the same distribution (signal), the probability of Hits [p(Hit)] and Misses [p(Miss)] sum to 1.0, as do the probabilities of False Alarms [p(FA)] and Correct Rejections [p(CR)]. Consequently, a complete description of the observer's behavior only requires information about p(Hit) and p(FA).

Sensitivity (d')

As noted above, the observer's ability to detect, or sensitivity, is determined in SDT by the distance between the means of the signal (μ_s) and noise (μ_n) distributions. This distance is formally measured in units of standard deviations (σ) by d':

$$d' = \frac{\mu_s - \mu_n}{\sigma}.$$

(5.1)

If the distributions are Gaussian and of equal variance, d' can be expressed as the difference of two standardized z-scores:

$$d' = z_s - z_n.$$

(5.2)

In practice, the values of μ_s, μ_n and σ are not known, and the probabilities p(Hit) and p(FA) are used to estimate z_s and z_n as noted in Wickens (2002, pp. 22–24). So Equation (5.2) can be rewritten as

$$d' = z_{\text{Hit}} - z_{\text{FA}}. \tag{5.3}$$

The values of z_{Hit} and z_{FA} are determined from p(Hit) and p(FA) by reference to standard tables of the normal or Gaussian distribution. When d' is zero, the observer has no sensitivity and is unable to distinguish between signal and noise. As d' increases, the distance between the means of the distributions increase, sensitivity increases and the observer is better able to detect signals.

The only aspect of SDT which is capable of altering detectability is the separation of the signal and noise distributions. In this regard, there are three options: decrease the level of background noise, increase the level of the signal, or change the variance of one or both distributions.

Bias, Decision-Making and Setting the Criterion

There are many different decision strategies that can be used to determine if an ambiguous event is a signal or noise. Some decision strategies may not even be rationally based. We will consider the most common strategy only. Decision strategies are usually the result of decision goals. A common goal in forming decisions is to maximize the expected value of the decision. Assume that the observer has a value, V, (positive or negative) for each of the outcome cells in Table 5.1. Table 5.2 illustrates this.

In Table 5.2 each of the outcomes also has a probability of occurrence. The expected value of an outcome, by definition, is its probability multiplied by its value. The probabilities of concern here are the joint probabilities of signal and a "Yes" response [$p(Y,s)$], a noise and a "Yes" response [$p(Y,n)$], etc. By definition, $p(Y,s)=p(\text{Hit}){\cdot}p(s)$, $p(Y,n)=p(\text{FA}){\cdot}p(n)$, etc. The expected value of the decision is the sum of all of the expected values for the outcomes. Hence the expected value, $E(V)$, for Table 5.2 is

$$E(V) = p(\text{Hit}) \cdot p(s) \cdot V(s,Y) + p(\text{FA}) \cdot p(n) \cdot V(n,Y)$$

$$+ p(\text{Miss}) \cdot p(s) \cdot V(s,N) + p(\text{CR}) \cdot p(n) \cdot V(n,N).$$

TABLE 5.2
Payoff matrix

		Observer Response	
		"Yes"	"No"
States of the world	Signal	$V(s,Y)$	$V(s,N)$
	Noise	$V(n,Y)$	$V(n,N)$

The goal is to maximize $E(V)$ which is accomplished by drawing the criterion line so as to achieve this. This is done by maximizing: $p(\text{Hit}) - \beta p(\text{FA})$. Green and Swets (1974) show that β is given by

$$\beta = \frac{V(n,N)+V(n,Y)}{V(s,N)+V(s,Y)}\,\frac{p(n)}{p(s)}.\tag{5.4}$$

If $p(s)=p(n)$, then β is only determined by the values of the outcomes. If all of the values of the outcomes are equal, β is only determined by the prior probabilities ($p(s)$ and $p(n)$). When the values of the outcomes are all equal and the prior probabilities are also equal, $\beta=1$. At the crossover of the signal and noise distributions in Figure 5.1, $\beta=1$.

Bias is the tendency of an observer to place his or her criterion anywhere except at the intersection of the noise and signal distributions (i.e., $\beta\neq 1$). Bias is independent of detectability (also called sensitivity or discriminability and measured by d' as noted above) and is determined by the observer's expectations (probability of signal, probability of noise), motivation (values of each of the decision outcomes), and other cognitive functions (e.g., memory, attention, decision strategy). If $\beta<1$, there is a bias for the observer to say "Yes, it is a signal" to an event. If $\beta>1$, there is a bias for the observer to say "No, it is not a signal". So, for instance, if the values of the decision outcomes in Table 5.2 are all equal, but $p(n)>p(s)$, then $\beta>1$ and there is a bias to respond "No". Likewise, if $p(n)<p(s)$, $\beta<1$ and there is a bias to respond "Yes". Keep in mind that the events in both situations are equally discriminable.

While changing bias does not affect d', it does change the probabilities of Hits and False Alarms. This can be seen in Figure 5.2 which plots changes in $p(\text{Hit})$ and $p(\text{FA})$ as the criterion is moved.

Each curve is a Receiver Operating Characteristic (ROC) curve. They show the covariance of $p(\text{Hit})$ and $p(\text{FA})$ for four values of d', including $d'=0$ as the criterion is moved. The line for $d'=0$ is often called the major diagonal and represents zero

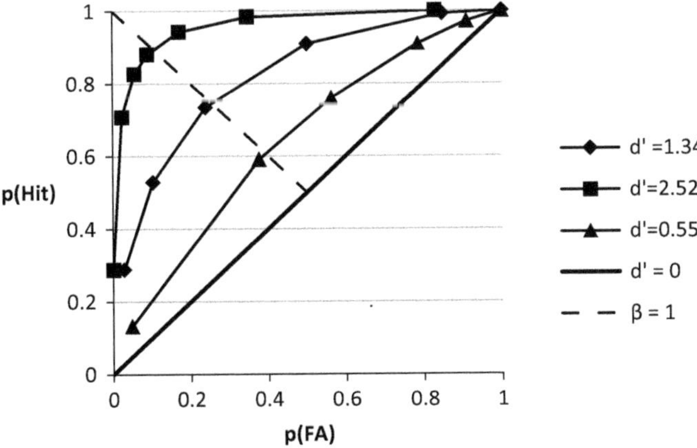

FIGURE 5.2 Receiver operating characteristic (ROC) curves showing that $p(\text{Hit})$ and $p(\text{FA})$ change as the criterion is moved. See text for details.

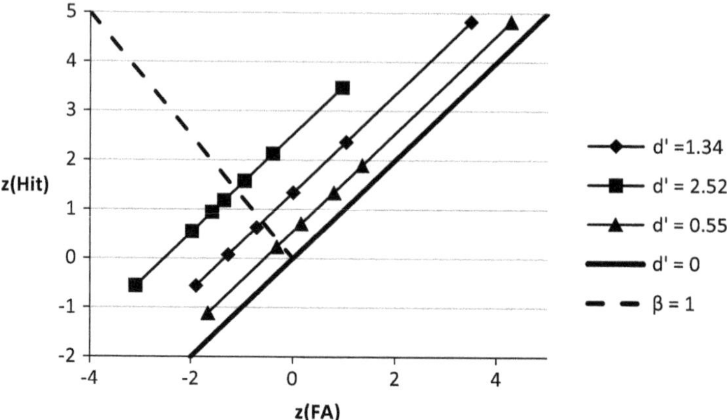

FIGURE 5.3   Data from Figure 5.2 plotted as z-scores.

sensitivity. The further a ROC is from the *major diagonal*, the higher the sensitivity of the observer. Figure 5.2 also shows the *minor diagonal* which is the dashed line labeled $\beta = 1$. This is the zero bias line. For points on a ROC above the minor diagonal, there is a bias to say "Yes" (i.e., $\beta < 1$). The further a point is from the minor diagonal, the greater the bias. p(Hit) and p(FA) increase because the criterion is moving to the left in Figure 5.1. There is a bias to say "No" for points on a ROC below the minor diagonal (i.e., $\beta > 1$). p(Hit) and p(FA) decrease for points further from the minor diagonal because the criterion is moving to the right in Figure 5.1.

It is usually assumed that the underlying distributions for noise and signal are normal or Gaussian and have equal variance. If the probabilities in Figure 5.2 are plotted as z-scores, this assumption can be tested because the ROC curves should be straight lines that are parallel to the major diagonal. This is shown in Figure 5.3 for the data in Figure 5.2.

β can also be calculated as the ratio of the ordinates of the standard normal curve corresponding to z(Hit) and z(FA):

$$\beta = \frac{y_{\text{Hit}}}{y_{\text{FA}}}, \tag{5.5}$$

where y_{Hit} is

$$y_{\text{Hit}} = \frac{1}{\sqrt{2\pi}} e^{\frac{-z(\text{Hit})^2}{2}}, \tag{5.6}$$

and y_{FA} is similarly defined.

Alternative Measures of Sensitivity and Bias

It is fairly easy in a laboratory setting to systematically vary sensitivity and bias so as to generate ROC curves and determine if the assumption of equal variance normal distributions for signal and noise are valid. This is much more difficult to do in a field setting, although it is worth the effort if the nature of the underlying distributions has practical or theoretical implications. In the event that it is highly doubtful that equal variance normal distributions are representative, indices of sensitivity and bias that are free of these parametric assumptions can be used. These indices are based on the geometry of the unit square corresponding to the ROC graph in Figure 5.2 (see Frey and Colliver, 1973 for more details). A' is the sensitivity measure and is defined as

$$A' = \frac{\left(p(\mathrm{Hit}) - p(FA)\right) + \left(p(\mathrm{Hit}) - p(\mathrm{FA})\right)^2}{4p(\mathrm{Hit})(1 - p(\mathrm{FA}))} + 0.5 \tag{5.7}$$

B'' is the bias measure and is defined as

$$B'' = \frac{\left(p(\mathrm{Hit}) - p(\mathrm{Hit})^2\right) - \left(p(\mathrm{FA}) - p(\mathrm{FA})^2\right)}{\left(p(\mathrm{Hit}) - p(\mathrm{Hit})^2\right) + \left(p(\mathrm{FA}) - p(\mathrm{FA})^2\right)}. \tag{5.8}$$

Both A' and B'' can be calculated when the probabilities are 1 or 0, unlike d' and β. A' can range from 0.5 (no sensitivity) to 1.0 (maximum sensitivity). B'' can range from -1.0 (bias to say "Yes") to 1.0 (bias to say "No").

ROC curves can be analyzed to determine the underlying probability distributions by using various transformations of $p(\mathrm{Hit})$ and $p(\mathrm{FA})$. For instance, if the underlying distributions are negative exponential, plotting $\log p(\mathrm{Hit})$ against $\log p(\mathrm{FA})$ will result in a series of straight lines with different slopes that converge at the upper right hand corner of the ROC graph (coordinates 1, 1). See Egan (1975) for a complete discussion of ROC analysis and the derivation of measures of sensitivity and bias for various probability distributions.

THEORY OF THE IDEAL OBSERVER

Ideal observers are created in SDT to model the performance of the perfect, or ideal, observer. Such an observer is maximally effective in using all of the available information to reach a rational decision within the bounds of the model limits. The behavior of real observers can then be compared with that of the ideal observer to determine if the model of behavior has validity in the search for underlying processes, or as a means to improve observer performance (Swets et al., 1964/1988).

Two examples of the use of Ideal Observers occur in the Track Inspection Time Study (Al-Nazer et al., 2011) and in a study to understand how different warning devices and countermeasures influenced drivers' decisions at grade crossings (Yeh et al., 2013).

In the track inspection time study, the real observer is hypothesized to depend on visual acuity and detection accuracy, visual search time, and attention (vigilance, distraction, fatigue) to detect defects in track that vary in size. Visual acuity determines the observer's ability to resolve detail. Visual acuity is affected by light intensity and exposure time. These variables can be used to determine a theoretical maximum speed (the observer is in a vehicle traveling on the track) at which objects of various sizes can be detected with a particular degree of accuracy. Visual search requires time and that time increases with the number of items in the search. Because visual search consumes time, it affects the theoretical maximum speed for detection at a particular degree of accuracy. Finally, if track inspection is considered a vigilance task, it is well known that detection accuracy declines as a function of time on task. Consequently, defects must be closer to the observer, or the speed of the observer must be reduced to maintain a desired level of accuracy.

In the grade crossing application, the real observer is hypothesized to use subjective estimates of the arrival time of the highway vehicle and of the train at the grade crossing to make decisions about stopping or proceeding through the grade crossing. Visual search for and localization of the train consumes time. A decision to stop can be made (or not) during this time and directly affects accident probability. Different warning devices and countermeasures such as reflectorization, train horns, and altering lights on trains all affect visual search and accident probability. For a particular situation there can be multiple ideal observers, so comparing the performance between real and ideal observers allows the mechanisms of detection and decision-making to be understood.

REFERENCES

Al-Nazer, L., Raslear, T., Patrick, C., Gertler, J., Choros, J., Gordon, J., & Marquis, B. (2011). *Track inspection time study* (Report No. DOT/FRA/ORD-11/15). Washington, DC: U.S. Department of Transportation.

Egan, J. P. (1975). *Signal detection theory and ROC analysis.* New York: Academic Press.

Frey, P. W., & Colliver, J. A. (1973). Sensitivity and responsivity measures for discrimination learning. *Learning and Motivation, 4,*327–342.

Green, D. M., & Swets, J. A. (1974). *Signal detection theory and psychophysics.* New York: Krieger.

Macmillan, N. A., & Creelman, C. D. (2005). *Detection theory: A user's guide* (2nd edn). Mahwah, NJ: Erlbaum.

McNicol, D. (1972). *A primer of signal detection theory.* London: Allen & Unwin.

Parzen, E. (1960). *Modern probability theory and its applications.* New York: Wiley.

Raslear, T. G. (1996). Driver behavior at rail-highway grade crossings: A signal detection theory analysis. In A. A. Carroll & J. L. Helser (Eds.), *Safety of highway-railroad grade crossings. Research Needs Workshop: Appendices* (Vol. II, Report No. DOT/FRA/ORD-95/14.2, DOT-VNTSC-FRA-95-12.2, pp. F9–F56). Washington, DC: U.S. Department of Transportation. https://www.fra.dot.gov/eLib/details/L04204#p12_z5_gD_kRaslear

Swets, J. A. (Ed.). (1964/1988). *Signal detection and recognition by human observers.* New York: Wiley (Reprinted by Peninsula Publishing, Los Altos, CA).

Swets, J. A. (1996). *Signal detection theory and ROC analysis in psychology and diagnostics: Collected papers.* Mahwah, NJ: Erlbaum.

Swets, J. A., & Pickett, R. M. (1982). *Evaluation of diagnostic systems: Methods from signal detection theory.* New York: Academic.

Swets, J. A., Tanner, W. P., & Birdsall, T. G. (1964/1988). Decision processes in perception. In J. A. Swets (Ed.), *Signal detection and recognition by human observers* (pp. 3–57). New York: Wiley (Reprinted by Peninsula Publishing, Los Altos, CA).

Wickens, T. D. (2002). *Elementary signal detection theory.* Oxford: Oxford University Press.

Yeh, M., Raslear, T., & Multer, J. (2013). *Understanding driver behavior at grade crossings through signal detection theory* (Report No. DOT/FRA/ORD-13/01). Washington, DC: U.S. Department of Transportation.

Section III

Grade Crossings and Trespassing

6 Grade Crossing Safety and Trespassing

Up until 1997 grade crossing accidents were the primary cause of fatalities in the railroad industry. In 1997 there were 461 grade crossing deaths and 533 trespasser deaths. Since 1997, trespasser deaths have remained the primary cause of fatalities in the industry (see Figure 6.1).

Moreover, while grade crossing deaths have declined since 1997, trespasser deaths have remained fairly constant. Grade crossing incidents and injuries have, likewise, declined since 1978, while trespassing incidents and injuries have been nearly constant over the same time period (see Figures 6.2 and 6.3). Trespassing incidents are very likely to result in either death or an injury. Roughly 50% of trespassing incidents end in death or injury. Grade crossing incidents, however, only result in 11% deaths and 39% injuries. Roughly 50% of grade crossing incidents do not have casualties.

GRADE CROSSING SAFETY

Motorist Behavior, Grade Crossing Safety, and Signal Detection Theory

Why do motorists drive into the sides of trains? This was a question I asked myself when I joined FRA in 1993 and began to educate myself concerning the safety issues in the railroad industry. It struck me as an important question because trains are very large visual stimuli that would be difficult to not see. Moreover, motorists also drive in

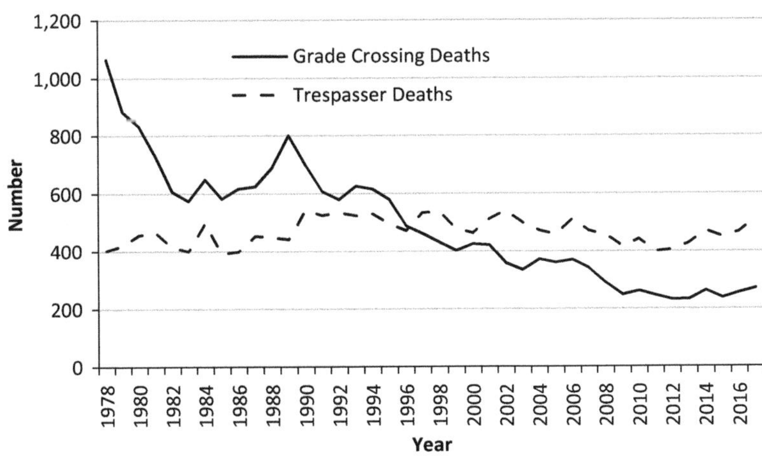

FIGURE 6.1 Grade crossing and trespasser deaths.

DOI: 10.1201/9781003500018-9

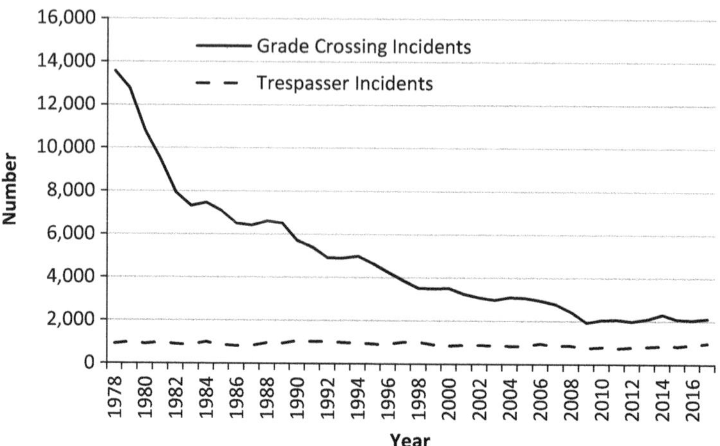

FIGURE 6.2   Grade crossing and trespassing incidents.

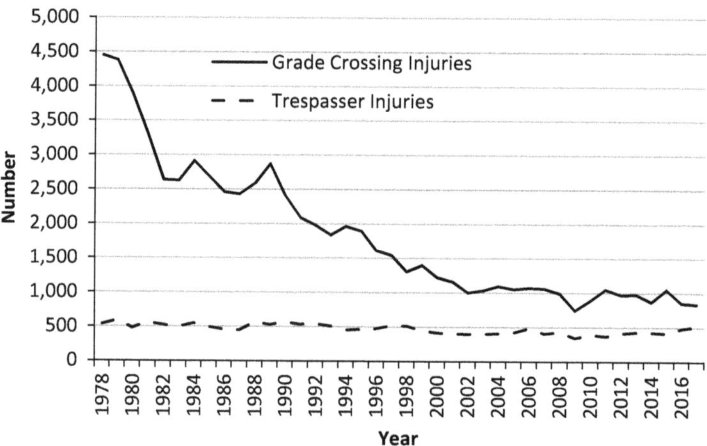

FIGURE 6.3   Grade crossing and trespassing injuries.

front of trains, even when the proximity of a train is indicated by the sounding of the train horn and the activation of bells, lights and gates at active grade crossings. Why?

"Research sponsored by the Federal Railroad Administration (FRA) has identified driver behavior as the main cause of highway-rail grade crossing crashes and that factors such as train and traffic volume can contribute to the risk of a crash." (United States Government Accountability Office, 2018, p. 1). This may seem obvious, but in 1996 I had to convince the then Associate Administrator for Railroad Development (Jim McQueen) that FRA grade crossing research should focus on motorist behavior. FRA shares responsibility with the Federal Highway Administration (FHWA) for grade crossing safety, but for FHWA the relatively small number of deaths and fatalities at grade crossings diminishes the importance of this highway safety risk. It is an interesting fact that the FHWA tends to refer to grade crossings as "Railroad-Highway

Grade Crossings" (see for example, FHWA, 1986), while the FRA tends to refer to "Highway-Railroad Grade Crossings" (see examples in FRA reports in reference list). This is not just about labeling, the labels reflect underlying differences in policy.

Although motorist behavior is now almost universally acknowledged to be a major cause of grade crossing incidents, the official policy of FRA is not guided by a comprehensive scientific theory of motorist behavior. As noted by the GAO (2018, p. 12), "...FRA has sponsored a number of research efforts to better understand the causes of grade-crossing crashes and identify potential ways to improve engineering, education, and enforcement efforts." Engineering, education, and enforcement, also known as the 3 Es, has guided FRA policy regarding grade crossing safety for more than three decades. It is, however, more of a slogan than a comprehensive scientific theory of motorist behavior. Consider, for example, the quote from the highlights of the GAO report: "...factors such as train and traffic volume can contribute to the risk of a crash." What do the 3 Es tell us about train traffic and crash risk? A scientific theory of motorist behavior should tell us, not only what the relationship is between train traffic and crash risk, but why that relationship exists. It should also be able to predict new aspects of motorist behavior that are empirically testable. A comprehensive theory of motorist behavior should be able to incorporate the 3 Es and predict what changes in engineering, education and enforcement are most likely to affect positive changes in grade crossing safety.

I credit the idea of using Signal Detection Theory (SDT) to model motorist behavior to Neil Lerner and associates (Lerner et al., 1990, pp. 3–12) who wrote: "A related principle from the area of signal detection theory is that the higher the perceived probability of an event, the higher is the likelihood that an observer will report having detected the event. If the driver assigns a low probability to the presence of a train at a rail-highway crossing, he will adopt a higher criterion for detecting the train, and this will increase his chances of missing the train. It is important to note that the criterion for detection is not consciously set, but rather corresponds to the amount of visual 'evidence' required for detection."

In other words, if a motorist has a low perceived probability of a train at a grade crossing, that motorist has increased probability of not detecting the train and, consequently, of having an accident. The prediction from SDT is that the frequency of trains should vary inversely with the accident rate at the crossing. This goes directly to the question, "What do the 3 Es tell us about train traffic and crash risk?" It seems counterintuitive that a higher frequency of trains should result in a lower accident rate. The 3 Es are mute on this point. However, as noted in Chapter 5, the prior probability of a signal (a train) is important in determining whether an observer has a bias to say "Yes, I detect a train", or "No, I do not detect a train". If $p(n) > p(s)$, then $\beta > 1$, and there is a bias to respond "No". If $p(n) < p(s)$, then $\beta < 1$, and there is a bias to respond "Yes". The value of $p(n)/p(s)$ sets the value of β, (all other factors being equal), so SDT predicts that as $p(s)$ increases, bias to say "Yes" increases, correct detections of trains increase (Hits) and accidents decrease (Misses). Assuming that train probability is a monotonic function of train frequency, it follows that the grade crossing accident rate should vary inversely with the frequency of trains at crossings. This relationship is shown in Figure 6.4 which is adapted from Raslear (1996).

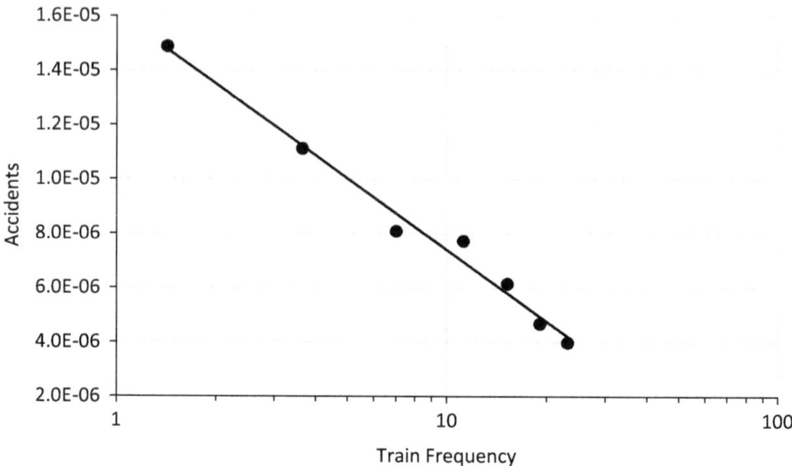

FIGURE 6.4   Accidents/crossing/year/train/car as a function of trains/day.

Each data point in Figure 6.4 comes from one of the seven different types of grade crossing devices in use in 1986. The accident rate is normalized for the number of grade crossing devices of each type, the average number of trains per year for each device type, and the average number of cars per year for each device type (see Chapter 4). Thus, Figure 6.4 shows the equal exposure accident rate for each device.

Translating Signal Detection Theory into a Model of Motorist Behavior

Why model motorist behavior? According to R. M. May (2004, p. 791), "The virtue of mathematics ... is that it forces clarity and precision upon ... conjecture, thus enabling meaningful comparison between the consequences of basic assumptions and the empirical facts." As noted above, the 3 Es do not do this. The development of a mathematical model of motorist behavior from SDT will, it is hoped, encourage a more scientific approach to grade crossing issues.

The basic terminology of SDT, as described in Chapter 5, can be confusing to someone not trained in Psychology and dealing with grade crossing accidents. A senior member of the FRA Office of Research and Development, Bob McCown, pointed this out to me when he reviewed the first draft of Raslear (1996). For instance, Table 5.1 in Chapter 5 designates a "Yes" response to a signal as a Hit and a "No" response to a signal as a Miss. If a motorist responds "No" when a train is present, an accident results. The motorist did not "miss" the train, the motorist "hit" the train or the train "hit" the motorist. Table 6.1 shows the orthodox SDT terminology and the terminology devised to clarify what happens at grade crossings.

In Table 6.1, if a train is close and the motorist decides not to stop, an Accident (AC) occurs. The decision to stop when a train is close is termed a Valid Stop (VS). The decision criterion divides the distribution of "train is close" percepts (signal distribution in Figure 5.1 of Chapter 5) into Valid Stops and Accidents. The criterion also divides the distribution of "train is not close" percepts (noise distribution in Figure 5.1) into two parts: False Stops (FSs) and Correct Crossings (CCs).

TABLE 6.1

Four Possible Outcomes in an SDT Matrix for a Motorist at a Grade Crossing

		Motorist Response	
		"Yes", stop	**"No", don't stop**
States of the world	Train close	*Hit/valid stop*	*Miss/accident*
	Train not close	*False alarm/false stop*	*Correct rejection/correct crossing*

Given clarified terminology, the next problem is how to estimate the key data necessary for an SDT analysis of grade crossings, namely $p(VS)$ and $p(FS)$. Raslear (1996) suggested that, since $p(AC) = 1 - p(VS)$, that the equal exposure accident rate at the various types of grade crossings was the way to estimate $p(VS)$. In this instance the equal exposure accident rate is accidents per crossing per car per train per minute.[1] This rate can be used in Equation (4.1) (Poisson probability law) from Chapter 4 to estimate $p(AC)$ and $p(VS)$.

Estimating $p(FS)$ is a bit more complicated. We start by observing that the risk of an accident (AR) can be defined as the probability that a car $p(car)$ and a train $p(train)$ will be in the grade crossing at the same time $[p(car) \times p(train) = p(AR)]$. Again, using Equation (4.1) from Chapter 4, the number of trains per minute and the number of cars per minute at various types of grade crossings provides an estimate of $p(car)$, $p(train)$ and $(p(AR))$. $p(AR)$ is not the same as $p(AC)$[2]. If $p(AR)$ is the probability that a car will not safely cross railroad tracks, then $1 - p(AR)$ is the probability that a car will cross without an accident (a correct crossing as in Table 6.1). Thus $1 - p(AR) = p(CC)$, and by definition $p(FS) = 1 - p(CC)$ so that $p(FS) = 1 - [1 - p(AR)]$. Consequently, $p(AR)$ is the estimator of $p(FS)$.

Given estimates of $p(VS)$ and $p(FS)$, values of d' and β can be calculated for the various grade crossing devices. Figure 6.5 (see Yeh et al., 2013) plots $z(VS)$ as a function of $z(FS)$ for 1986 and 2007 (see Chapter on SDT regarding the z-transform of probabilities). Examination of Figure 6.5 shows that equal variance Normal distributions are good representations of the signal and noise distributions since the dotted lines drawn through the points for each year (set at mean d' values for each year) are straight lines with unit slope, parallel to each other and the line for $d' - 0$. The value of d' has reliably increased from 1986 to 2007 (from 6.95 to 7.24, a 4% change). The solid diagonal extending from the top left of Figure 6.5 to point (0, 0)) is the zero-bias line. A value of $\beta < 1$ indicates a bias to stop (the area above the diagonal), and a value of $\beta > 1$ indicates a bias to proceed (the area below the diagonal). It is clear from Figure 6.5 that β has substantially changed from 1986 to 2007. In 1986 mean β was 1.45, indicating that motorists were more likely to proceed than to stop at grade crossings. In 2007 mean β was 0.024 (a 60-fold change), indicating that motorists were more likely to stop than proceed. This conservative change occurred for all warning devices and was statistically reliable.

Figure 6.5, and analyses by Raslear (1996) and Yeh et al. (2013) of device effectiveness (defined as the ratio, $p(FS)/p(AC)$) and changes in d' and β, indicate that motorist decision making (setting the criterion) is the most important factor in grade crossing safety. Essentially this means that grade crossing devices improve safety

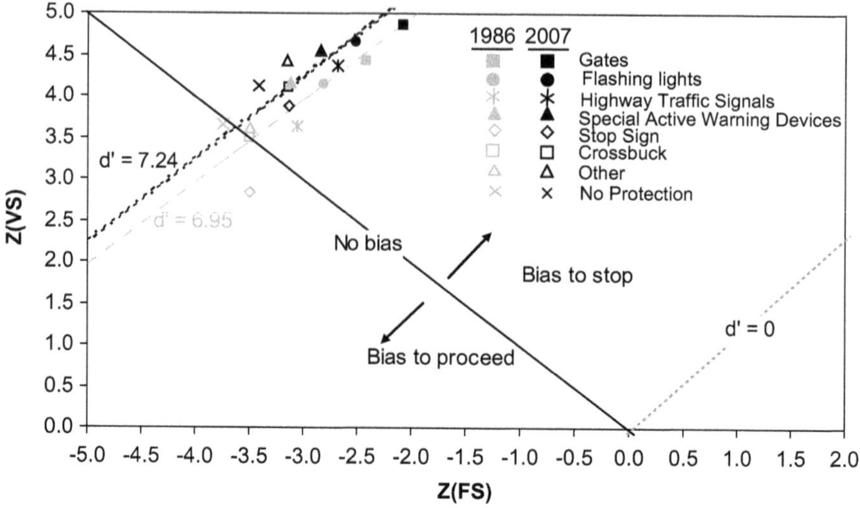

FIGURE 6.5 z(VS) as a function of z(FS) for 1986 and 2007. (From Yeh et al., 2013.)

because they do more to increase motorists' bias to stop (β) than by increasing the signal-to-noise ratio (d').

A key question for translating SDT into a model of motorist behavior at the grade crossing is "what is the basis for a decision that the train is close?" The percept, "train is close", needs to be a physical entity that drives the decision to stop or proceed across the grade crossing. In essence the model of motorist behavior is a Theory of the Ideal Observer for motorist behavior (see Chapter 5). In this instance, Raslear (1996, pp. 34–40) hypothesized that the motorist bases decisions on subjective estimates of arrival time of the train and the highway vehicle at the grade crossing. Before these arrival times can be estimated, the train must be visually searched for and localized. These processes consume time and affect accident probability.

The sight distance (d_h) determines the physical time (T_s) required to safely stop the highway vehicle without encroaching the crossing area. d_h depends on the distance of the motor vehicle from the tracks, the velocity of the motor vehicle, and several other factors. The sight distance (d_t) determines the physical time (T_c) required for the motor vehicle to cross and be clear of the crossing when the train arrives at the crossing (see Tustin et al., 1986, p. 132). The motorist's perception of these times, however, is not veridical. It is typically found in the human time perception literature that short durations are overestimated and that long durations are underestimated. Kaiser and Mowafy's (1993, Figures 7 and 8) estimate of the relationship between a judged time (T^*) and an actual time (T) were used to adjust the values of T_c and T_s to reflect this aspect of human time perception: $T^* = 0.84375\,T + 0.84375$.

For instance, using the relationship above, a 4 s duration would be judged to take 4.22 s and an 8 s duration would be judged to take 7.59 s.

The motorist decides to stop or to cross on the basis of the perceived difference between judged T_c (T_c^*) and judged T_s (T_s^*). Raslear (1996) modeled the perceived difference between the two durations as

$$d' = \frac{\mu\left(T_C^* - T_S^*\right)}{\sigma\left(\left(T_C^*\right)^2 + \left(T_S^*\right)^2\right)^{0.5}},$$ (6.1)

where $\mu/\sigma = \gamma$ is the inverse of the Weber constant for time. A study of perceived time to collision by Bootsma and Oudejans (1993, experiment 1) estimated that $\gamma = 57.47$.

In Chapter 5, it was noted that in the Gaussian or Normal distribution SDT model, the mean and the variance are independent. This is not the case in Equation (6.1). The model in Equation (6.1) is based on a widely applied theory of time discrimination (Gibbon, 1977), Scalar Expectancy Theory (SET). SET assumes that physical durations generate distributions of time estimates in which the mean and the standard deviation of the distribution are proportional to the interval being timed (T), so that the coefficient of variation is a constant:

$$\text{Mean} = \mu T,$$ (6.2)

$$\text{Standard deviation} = \sigma T,$$ (6.3)

$$\text{Coefficient of variation} = \gamma = \mu/\sigma.$$ (6.4)

This is reflected in Equation (6.1) and all subsequent equations concerning the calculation of d'.

Because time perception is being used to model driver behavior at the grade crossing, Yeh et al. (2013) used an information processing heuristic to provide a psychological context for changes in detection and bias. Figure 6.6 is a modification of that heuristic model of time perception.

T_s^* and T_c^*, are discriminated in the following fashion. The sight distances, d_h and d_t are detected in the *Sensory Function*. This sensory information is gated by a

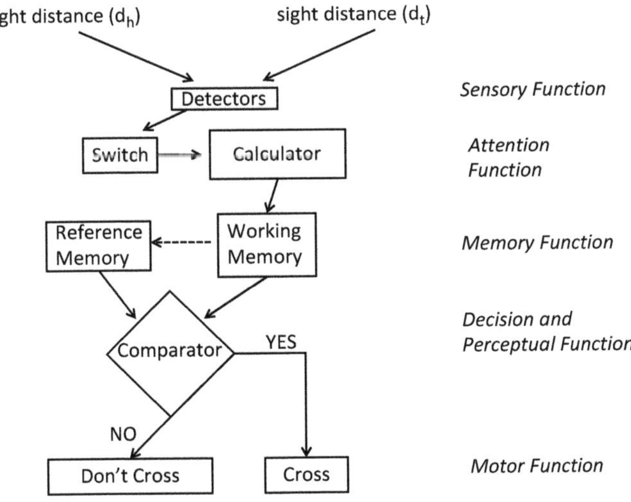

FIGURE 6.6 Heuristic information processing model of time perception.

Switch which is the *Attention Function*. The Switch has a latency to close and open and mimics the other commonly observed phenomena of attention. There is inherent variability in the Switch, so that sensory representations of d_h and d_t are inaccurate when passed to the Calculator. The Calculator uses d_h and d_t to estimate T_s^* and T_c^*. As a result, the Calculator outputs are inaccurate indicants of T_s^* and T_c^*. The Calculator outputs are stored in Memory (the *Memory Function*) where they can be compared with a reference value or criterion. The reference value is adjusted by the outcomes of the observer's decisions. Both the memory of the Calculator output and the reference value are imperfect and increase the variance of the system. Decisions are based on the comparison of the memory of the Calculator output and the reference value by a decision rule. A common decision rule, and the one assumed here, is that decisions are made to maximize expected value of the decision outcomes.

Attention and memory change the denominator of Equation (6.1) because these functions are assumed to increase variance, which acts to decrease d'. Switch variance (δ) is usually assumed to be a constant, independent of the durations judged, which is added to the total variance (see Church and Meck, 1984):

$$d' = \frac{\mu\left(T_C^* - T_S^*\right)}{\sigma\left(\left(T_C^*\right)^2 + \left(T_S^*\right)^2 + \delta\right)^{0.5}}, \tag{6.5}$$

where $\delta = 5.21$ based on Creelman (1962). Memory variance (ω), on the other hand, is assumed to be a function of the duration judged (Creelman, 1962). So a generic ideal observer with both memory and switch functions would be expressed as

$$d' = \frac{1}{\sqrt{\omega}} \frac{\mu\left(T_C^* - T_S^*\right)}{\sigma\left(\left(T_C^*\right)^2 + \left(T_S^*\right)^2 + \delta\right)^{0.5}}. \tag{6.6}$$

Estimates of ω in the literature for time discriminations at grade crossings do not appear to be reasonable. Since memory variance decreases the value of d' with time, an exponential time decay function, which is also suggested for bias (see Equation 6.9 below), seems appropriate. Because memory decays in real time the decay function depends on real time (T_c and T_s rather than T_c^* and T_s^*). Thus, sensitivity is expressed as

$$d' = ke^{-\lambda T} \frac{\mu\left(T_C^* - T_S^*\right)}{\sigma\left(\left(T_C^*\right)^2 + \left(T_S^*\right)^2 + \delta\right)^{0.5}}, \tag{6.7}$$

where k is a constant, λ is the decay rate, $\delta = 5.21$, and $\mu/\sigma = \gamma = 57.47$.

Yeh et al. (2013) used a version of Equation (6.7) to model an Ideal Observer's sensitivity for grade crossing countermeasures designed to improve detection of the train which could be compared to empirical measures of d'. They used a version of Equation (6.12, below) to model an Ideal Observer's bias for the same countermeasures for comparison to empirical measures of β. The countermeasures included: Commercial Motor Vehicle Safety Measures, Alerting Lights on Locomotives, Reflectorization of

Rail Cars, Improving Sight Lines, and Improving Grade Crossing Warning Reliability. A complete comparison of estimated and empirical sensitivity and bias can be found in Yeh et al. The reader should remember that it is expected that an ideal observer performs better than a real observer because the ideal observer uses all the available information to the fullest extent possible. Below we briefly outline their results.

Sensitivity (Ability to Detect the Train)

With the exception of Commercial Motor Vehicle Safety Measures and Grade Crossing Warning Reliability, all the other countermeasures improve detection through visual search. Yeh et al. develop the ideal observer from a model in which visual search for the train is aided by the train horn. This proceeds to sight line improvements through the removal of visual clutter at the grade crossing, the addition of alerting lights, and the reflectorization of rail cars to further reduce visual search time and improve train detection.

Yeh et al. assumed that a motorist is approaching the grade crossing at a speed of 30 mph, and that a train is approaching at 10 mph. A visual field of approximately 180° must be scanned, and the time required to detect the train increases as the number of items in the visual field increases. Luce (1986) estimated the impact of clutter on visual search. An estimate of the average time to search for a target can be derived from the following formula:

$$S = \tfrac{1}{2}kM + r_0, \tag{6.8}$$

where k = mean time required to search an item, M = the number of items, and r_0 = residual time.

The variance in the visual search can be described by

$$\eta^2 = \eta_m^2 M + \eta_r^2, \tag{6.9}$$

where η_m^2 = per item variance, and η_r^2 = residual variance. Values for these parameters using data from the human factors literature are as follows: $k = 0.02$ (Sternberg, 1966), $r_0 = 0.4$ (Sternberg, 1966), $\eta_m^2 M = M^2/12$, and $\eta_r^2 = 8.2944$ (Luce, 1986). The effect of visual clutter was then determined by adjusting the values for the perceived time to cross, taking into account sight lines and the perceived time to stop by the value of S, and accounting for the variance, σ, in the calculation of d'.

Visual search at the grade crossing may be aided by the sound of the train horn. Data regarding sound localization indicates that pure tones are localized with a 10-degree error. Therefore, the sound from a train horn reduces the search field to 10° (i.e., by one-eighteenth). This reduces S which affects the perceived times to cross and to stop. It also reduces the variance due to visual search by a factor of 18. The use of a train horn, as can be seen in Figure 6.7, does reduce accident risk, as was previously demonstrated theoretically by Raslear (1996), and was empirically documented by the FRA (1995).

Carroll et al. (1995), quantified the potential benefit of alerting lights by comparing the distances at which trains could be detected when equipped with different visual alerting devices. Trains equipped with crossing lights could be detected from a distance of 1,548 ft and the standard headlight alone was detected from a distance of 1,257 ft.

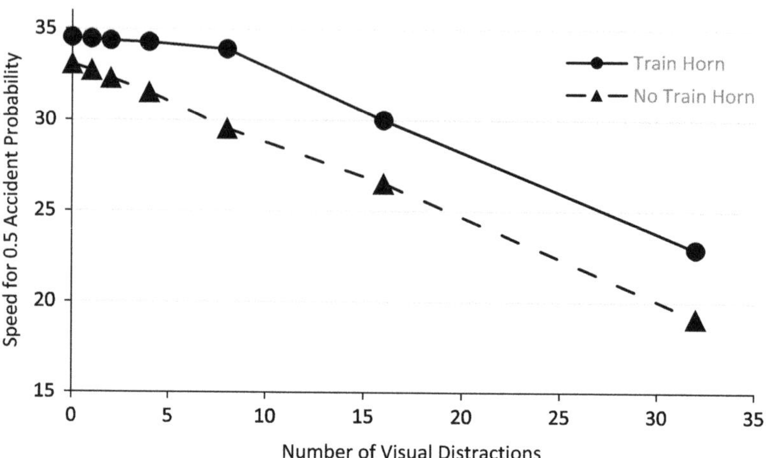

FIGURE 6.7   Effect of train horn on accident risk as a function of visual clutter. (Adapted from Raslear, 1996.)

For a train travelling 25 mph, a train with crossing lights can be detected 42.2 seconds before its arrival, whereas a train with headlight alone can be detected 34.3 seconds before arriving at the crossing. A 23% increase in T_c and T_s in the baseline model above.

Similarly, a study by Multer et al. (2001) was used to determine the effect of refectorization. They found that trains with reflectors were detected at almost three times the distance of trains without reflectors (1,113 ft for reflectorized trains, 373.5 ft for trains without reflectors). Thus, reflectors were estimated to improve detection distance 2.97 times that for trains without reflectors. However, not all trains had reflectors in 2007. Yeh et al. assumed the implementation rate at 60%.

Therefore, T_{s+} (change in T_s due to alerting lights and reflectors) was calculated as:

$$T_{s+} = (T_s \times 1.23 \times 2.97 \times 0.6) - (S/18). \tag{6.10}$$

The benefit of reflectors is primarily in reducing accidents in which a driver runs into the train (RIT). Thus the impact of reflectors is expected when the train is *at* the crossing. Therefore,

$$T_{c+} = (T_c \times 1.23) - (S/18). \tag{6.11}$$

Yeh et al. found that adding alerting lights and reflectors increased d' for 61 items of clutter from 6.95 in the baseline model to 11.22. Improving sight lines until there are only 4 items left at the grade crossing, further increases d' to 12.11. The average sensitivity for 2007 was 7.24.

Commercial Motor Vehicle (CMV) rules do not simply affect visual search times and variance, they change driver behavior. The rules require commercial drivers to stop at all crossings. The decision faced by the CMV driver is not whether to stop or cross, it is whether to cross following a stop. Thus, an estimate of sensitivity needs only to consider train speed because the highway vehicle speed is zero.

If train speed is 10 mph (14.67 ft/s), $T_c = 16.36$ s, the amount of time necessary for the highway vehicle to just clear the tracks from a stop, and the time for the train to reach the grade crossing (see Ogden, 2007, p. 68). Adjusting for alerting lights, reflectors, visual search with four items of clutter, and assuming that commercial drivers perform a "rolling stop" at the grade crossing instead of a full stop, $d' = 8.13$. Since the observed value of d' for CMV rules in 2007 was 7.52, the "rolling stop" scenario is reasonable.

Reliability is a serious problem when a warning device fails to activate. Motorists who experience a device activation failure may proceed with greater caution at grade crossings with active devices, crossing after a rolling stop, similarly to CMV drivers. A weighted average (based on exposure to warning device failure) of T_c, T_c^* and T_s^* values for CMV and for the combination of train horn, alerting lights, reflectorization, and improved sight lines (four items of visual clutter) allows d' to be calculated as $d' = 12.72$. This value of d' is higher than the mean for 2007, 7.24.

Bias (Willingness to Stop)

Bias is defined by Equation (6.4) in Chapter 5. Using the terminology developed above in this chapter, Equation (6.4) can be rewritten as

$$\beta = \frac{V(\text{FS}) + V(\text{CC})}{V(\text{VS}) + V(\text{AC})} \frac{p(n)}{p(s)}. \tag{6.12}$$

Since we assume that the driver's goal is to maximize expected value, the ideal observer would stop when the value of $\beta < 1$ and would cross when the value of $\beta > 1$. The effects of memory and attention have not been extensively explored in the literature on SDT. However, they are generally thought to affect both sensitivity and bias (see MacMillan and Creelman, 2005; Swets, 1996). Equation (6.12) is the ratio of the expected values of the decision outcomes associated with noise and stimulus. In economic theory, values decay with time. For instance, the present value of a good is reduced the further it is realized in the future. This idea is incorporated in Equation (6.13) which suggests that the time delays associated with T_s and T_n affect bias through the exponential decay function,

$$ke^{(-\lambda T)}, \tag{6.13}$$

where λ is the decay rate and k is a constant. The bias equation then is written as

$$\beta = ke^{(-\lambda T)} \frac{V(\text{FS}) + V(\text{CC})}{V(\text{VS}) + V(\text{AC})} \frac{p(n)}{p(s)}, \tag{6.14}$$

where $k = 0.9$, $\lambda = 0.001$, $p(s)$ is the probability of a train, $p(n) = 1 - p(s)$, and $V(\text{CC})$, $V(\text{FS})$, $V(\text{VS})$, $V(\text{AC})$ are the values associated with a payoff matrix that reflect the driver's motivation with respect to correct crossings, false stops, valid stops, and accidents, respectively. The term $[V(\text{CC}) + V(\text{FS})]$ is the value of noise ($V(n)$) whereas $[V(\text{VS}) + V(\text{AC})]$ is the value of signal ($V(s)$). In Equation (6.14), T is the subjective time T^* because subjective value decreases in subjective time.

Yeh et al. note that train traffic doubled from 1986 through 2007 from an average of 4.19 trains per day to 9.15 trains per day. This increases $p(s)$, decreases $p(n)$, and changes bias. The probability of a train ($p(s)$) at a grade crossing is estimated using the Poisson distribution, represented in Equation (6.15):

$$p(s) = 1 - e^{-\lambda_T}, \tag{6.15}$$

where λ_T represents the mean frequency of a train at a grade crossing in a 1-minute period of time. λ_T is calculated from the train rate per day at a crossing (see Raslear, 1996).

Accordingly, in 1986 $p(s)$ was 0.0029 and in 2007 doubled to 0.0058. As a result, the ratio $p(n)/p(s)$, also called the prior odds, was just a little over two times higher in 1986 (343.73) than in 2007 (170.63). The change in the *observed* bias for these years was much greater, however. In 1986, Yeh et al. estimated β at 1.45 and in 2007 the estimate was 0.02. Therefore, the change in *observed bias* was on the order of 70 times, which suggests that other factors contributed to the change in bias. Logically, the payoff matrix may account for these other factors. This is explored below for the specific safety countermeasures.

Yeh et al. estimated the personal expected economic impact of specific counter-measures through the payoff matrix. Commercial motor vehicle driver regulations disqualify a driver for a period of 60 days (d) up to 1 year if he or she commits a violation(s) at a grade crossing (see 49 CFR Part 383.51). A driver who commits three violations must be disqualified for one-year. A one-year disqualification would cost \$51,903 in lost wages. Horton et al. (2009) found that 23.7% of all grade cross-ing incidents in 2003 involved CMV drivers. Consequently, the maximum expected value of a grade crossing incident, based on violations of the CMV rule, would be $0.237 \times \$51,930$. Therefore,

$$\beta = 170.63 \times 0.9e^{-0.0017T_c^*} \times \left[\frac{1}{0.237 \times 51930} \right] = 0.0123, \tag{6.16}$$

where $T_c^* = 17.48$, and 170.63 is the prior odds. The observed value of β was 0.008 in 2007 and 0.562 in 1986 for CMV driver safety rules. Note that the subjective cost estimate to stop only captures lost wages. Fines or points added to the driver's per-sonal license could further reduce bias.

Since there are no documented costs and benefits associated with the other safety factors, Yeh et al. derived estimates for costs and benefits using Equation (6.14). $V(n)/V(s)$ for 1986 can be calculated given the known values of β_{1986} and the prior odds ($p(n)/p(s) = 343.73$). $\beta_{1986} < \beta_{2007}$ for all safety factors, so it follows that $V(s)$ has increased between 1986 and 2007. To estimate the change in $V(s)$ from 1986 to 2007, Yeh et al. used the funding provided to Operation Lifesaver as a proxy. The Department of Transportation's funding of Operation Lifesaver started in 1988 with \$69,000. In 2007, funding was \$1,797,328, so $V(s)$ increased 26.05-fold.

Table 6.2 shows the values of $\beta_{1986}, \left(\frac{V(n)}{V(s)} \right)_{1986}, \beta_{2007}, T_c^*$ for alerting lights, reflec-tors, sight lines and warning reliability. From these values, Yeh et al. calculated $\hat{\beta}_{2007E}$ for each safety factor using the following equation:

TABLE 6.2

Ideal Observer Estimates of Bias for 2007 ($\hat{\beta}_{2007E}$)

Safety Factor	β_{1986}	$\left(\dfrac{V(n)}{V(s)}\right)_{1986}$	β_{2007}	T_c^*	$\hat{\beta}_{2007E}$
Alerting lights	0.507	0.001475	0.006	7.84	0.0086
Reflectorization	0.188	0.000547	0.003	6.53	0.0032
Sight lines	0.142	0.000413	0.002	6.18	0.0024
Warning reliability	0.001	0.00000291	0.000092	7.42	0.00002
CMV	0.562	0.0016345	0.008	17.48	0.0123

$$\hat{\beta}_{2007E} = 170.63 \times 0.9e^{-0.001T_c^*} \times \left[\left(\frac{V(n)}{V(s)}\right)_{1986} \times \frac{1}{26.05}\right]. \tag{6.17}$$

For completeness, the parameters for CMV have been added to the table.

The correlation between $\hat{\beta}_{2007E}$ and β_{2007} is 0.9943 ($p < 0.01$).

Conclusion

Signal Detection Theory has obvious value as a model of human behavior at grade crossings. Both sensitivity and bias play a role in determining whether a motorist decides to stop at a crossing or to drive through. Both the empirical data and the models, however, show that bias plays a much larger role in motorist decision-making. Between 1986 and 2007, β empirically changed 6,000%. By comparison, d' changed 4%. The implication is that interventions to improve grade crossing safety should focus on changing bias (willingness to stop), rather than sensitivity (ability to detect the train). Although not discussed above, Yeh et al. modeled the effect of combining education and enforcement on bias. They estimated that $\beta = 0.00002$ with only education and that $\beta = 0.00001$ with both education and enforcement.

The modeling presented here should not be considered definitive. Other definitions of the physically controlling variables should be tested to determine if a better approach is available. Also, it should be noted that the basic SDT framework presented here is not the only form of SDT available. For instance, Parasuraman, Masalonis and Hancock (2000) have developed Fuzzy SDT.

SIGNAL DETECTION MODELS AS PREDICTIVE TOOLS

To this point SDT models have been discussed with regard to the analysis of existing data using assumptions concerning controlling variables. Such analyses are post hoc and as such are inferior tests of the model. In science, models are tested and refined by predicting previously unobserved phenomena. A theory of motorist behavior at grade crossings, such as SDT, provides a means to critically examine facts and inchoate hypotheses so that they can be more formally stated and vigorously tested. Theory allows facts to be organized in a meaningful and insightful way that expands understanding of the issues and leads to the generation of new hypotheses.

Motorist expectations of long waiting times at grade crossings is frequently used as a hypothesis to account for motorist noncompliance at grade crossings (Yeh et al., 2013). Raslear (2015) proposed a model of motorist behavior at grade crossings which explains how long motorists will stop at a grade crossing before driving across the tracks. Most grade crossing collisions occur when motorists drive around gates or proceed across railroad tracks when flashing lights signal an approaching train. Train arrival time as well as the overall duration of the warning are often cited as contributing to motorist noncompliance (see Yeh and Multer, 2008 for details). Although the Manual of Uniform Traffic Control Devices (MUTCD) and FRA regulations require a minimal advance warning time of 20 s, warning times vary considerably because the track circuits that activate crossing gates and other warning signals are often at a fixed distance from the grade crossing. Variations in train speed cause variability in warning times and increases in warning times have been linked to increases in grade crossing violations.

Richards and Heathington (1990) found that motorists expect a train within 20 s of warning device onset. 95% of motorists stop and wait within 10 s of train arrival, more than 50% stop and wait within 10–20 s of train arrival, and 30% stop and wait within more than 20 s of train arrival. The variability in warning time is often cited as the main contributor to warning times which are longer than motorists' expectations. Variability in train arrival time has been viewed as a likely cause of motorist non-compliance *per se*. Ogden (2007, p. 128) notes that "Reasonable and consistent warning times reinforce system credibility. Unreasonable or inconsistent warning times may encourage undesirable driver behavior." Jenness et al. (2006, p. 130) further state that "Drivers must perceive train warning information to be accurate and reliable. Predictions of train arrival time should be as accurate as possible." This, however, conflates the variability (variance, σ^2) of the time that the train arrives with the train's mean time to arrival (μ). It is not clear whether μ, σ^2 or both contribute to motorist noncompliance or to what degree. Motorist expectations about train arrival could depend on μ, σ^2 regardless of μ, or σ^2 relative to μ. Raslear (2015) directly explores that issue through the SDT model.

The basic SDT model for motorist expectation is Equation (6.12), in which β is determined by the prior odds and the value function of the decision outcomes. Raslear considered the prior odds to be the principal determinant of motorist expectation and set the value function equal to 1. The *subjective* probability of a train in the grade crossing can be represented by a Gaussian-like function as shown in Figure 6.8. Figure 6.8 has the familiar bell-shaped form of a normal probability density function (PDF), but is not a PDF since its integral is greater than 1 (e.g., the value of p(train) at 20 s is 1). The equation for this function is

$$p(\text{train}) = \exp\left[-\frac{(t - \text{MTA})^2}{2\sigma^2}\right], \tag{6.18}$$

where t is the arrival time of the train, MTA is the mean arrival time of the train, and σ^2 is the variance in the arrival time of the train. Equation (6.18) differs from the Gaussian PDF by a factor of $\dfrac{1}{\sigma(2\pi)^{0.5}}$. μ in Figure 6.8 is 20 s, and $\sigma = 7$ s.

β, based on the function in Figure 6.8 and Equation (6.12), is shown in Figure 6.9.

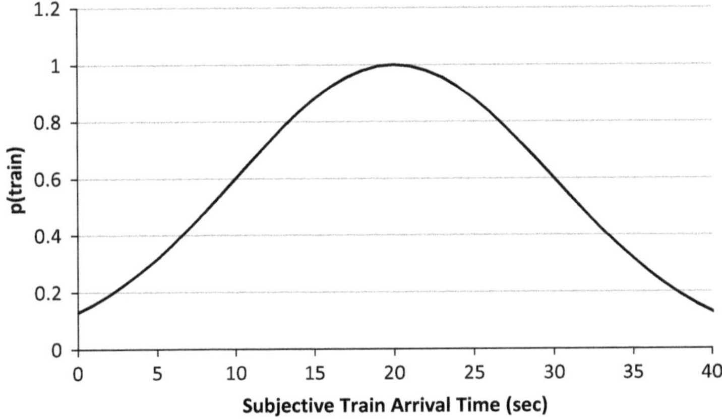

FIGURE 6.8   Subjective probability of a train in the grade crossing as a function of expected train arrival time. (After (Raslear, 2015.)

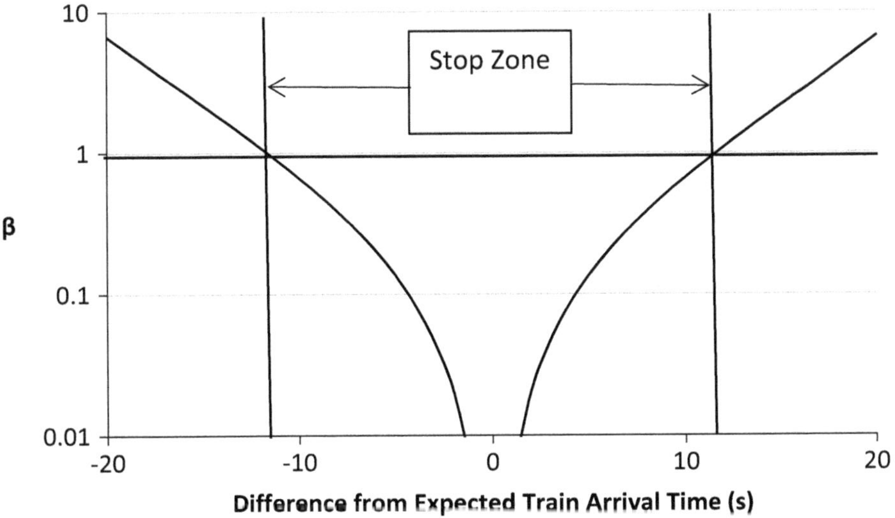

FIGURE 6.9 β as a function of difference from expected train arrival time. The stop zone indicates the time interval during which motorists are likely to stop. (After Raslear, 2015.)

When $\beta > 1$, there is a bias to proceed. When $\beta < 1$, there is a bias to stop. At $t = -20\,\text{s}$, warning signals at active grade crossings are activated. Figure 6.9 shows that there is a bias to proceed at $t < -11.5\,\text{s}$ (8.5 s after the warning signals are activated) and at $t > 11.5\,\text{s}$ (31.5 s after the warning signals are activated). In other words, the Stop Zone for $\mu = 20\,\text{s}$, $\sigma = 7\,\text{s}$ is 8.5–31.5 s. The greatest bias to stop is at μ $(t = 0)$. Raslear modeled β and calculated the stop zone for values of $\mu = 20$, 40 and 60 s and values of $\sigma = 1.25$, 10 and 30 s.

Figure 6.10 shows that σ accounts for almost 100% of the variance in the Stop Zone width. The width of the Stop Zone at each σ is the same regardless of the value of μ. This is consistent with the property of the Gaussian distribution that the mean is independent of the standard deviation.

Figure 6.11 shows that the calculations of p(stop), or compliance, based on a Gaussian model of motorist expectation of train arrival ($\mu = 20\,$s, $\sigma = 7\,$s) and high train detectability ($d' = 7.14$), approximates the Richards and Heathington data very well ($r^2 = 0.9959$) even though no attempt was made to specifically fit the data. A χ^2 test for goodness of fit (Hays, 1963) indicated that the Gaussian model was an adequate fit to the data ($\chi^2 = 5.89$, df $= 2$, critical value $= 5.99$, $p > 0.05$).

Raslear examined three probability distributions (Gaussian, χ^2, Poisson) to describe motorist stopping behavior, A Gaussian model provided the best fit to the Richards and Heathington data. In the Gaussian model mean train arrival time is

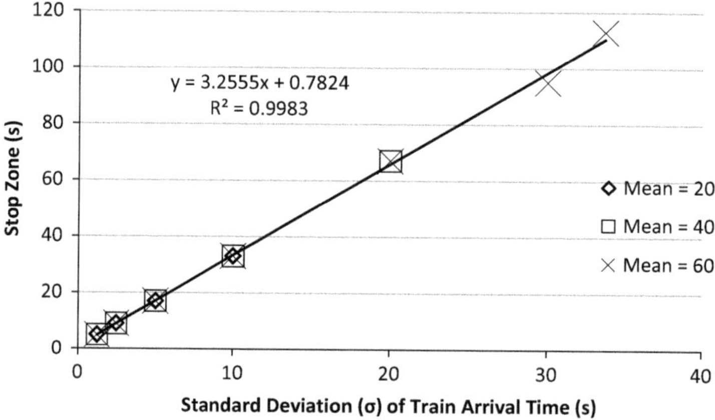

FIGURE 6.10 The stop zone width as a function of μ and σ. (After Raslear, 2015.)

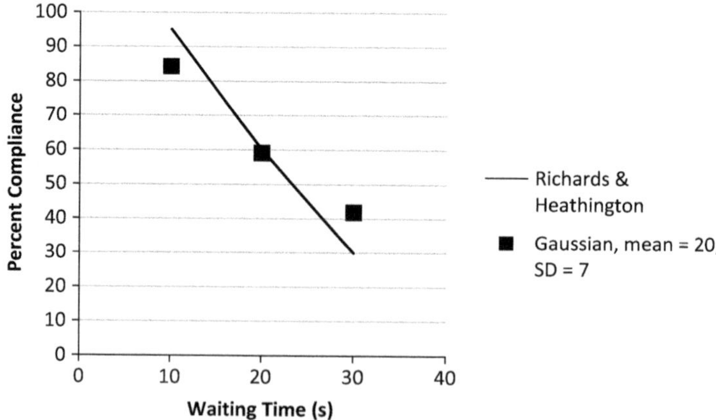

FIGURE 6.11 Percent motorist compliance as a function of waiting time. The data of Richards and Heathington are modeled by a Gaussian function. (After Raslear, 2015.)

not important, but train arrival time variance determines stopping behavior (see Figure 6.10) in that the stop zone increases with increases in the train arrival time variance. This seems counter-intuitive since variability of train arrival time at grade crossings is often thought to be the cause of motorist non-compliance. It is implicit in the grade crossing literature that increased variability in train arrival time, *per se*, (Jenness et al., 2006; Ogden, 2007) results in undesirable motorist behavior (e.g., non-compliance). However, the model makes sense if motorists are sensitive to uncertainty regarding train arrival. Under conditions with greater uncertainty about the arrival time of the train (higher variance or variability), it makes sense for a motorist to remain stopped. If the model is confirmed, engineering research efforts should be funded to maximize train arrival time variance. At this time, however, we lack sufficient empirical information about motorist behavior, mean train arrival time, and train arrival time variance. Three data points from one study do not constitute a solid empirical description of motorist behavior.

AUTOMATION AND SYSTEMS DESIGN

Many of the improvements in grade crossing safety through systems design and automation have already been mentioned in the text above. FRA devoted considerable resources to determining that train horns were sufficiently loud for motorists in cars with windows closed and radios on to detect the horn while not causing hearing damage to the locomotive crew and noise disturbance to the community surrounding the grade crossing (Keller and Rickley, 1993; Rapoza et al., 1999; Rapoza and Raslear, 2001; Rapoza and Fleming, 2002). Among suggestions to satisfy the inherent conflict of goals to promote motorist safety, protect crew hearing and minimize community noise, FRA also investigated the feasibility of replacing train horns with horns that would be deployed at the grade crossing (Multer and Repoza, 1998).

Reflectorization of rail cars and locomotives were also extensively researched with FRA funding. There were many issues involved, including whether specific patterns and colors of reflectors on rail cars improved the conspicuity of the rail cars, the durability and cost of the materials used, and the potential for accident reduction (Multer et al., 2001; Carroll et al., 1999; Ford et al., 1998). Based on this research FRA published a rule (Federal Railroad Administration, 2005) mandating the reflectorization of rail cars in the U.S. fleet. The cost/benefit analysis of the regulation employed "...a signal detection model consisting of an analysis of the statistical probability of different potential severities of hazard or injury and based on both laboratory experiments and data from FRA's RAIRS database..." (p. 169). In addition to reflectors, locomotive conspicuity was also increased by the use of auxiliary alerting lights (Carroll et al., 1995).

Photo enforcement at grade crossings started as a pilot project in the Los Angeles area in 1992 and ended in 1993 (Meadow, 1994). During that period of time there was an 84% reduction in grade crossing violations at some intersections. Carroll and Warren (2002) examined the outcome of photo enforcement projects at six highway-rail crossings across the United States. There was a reduction in violations ranging between 34% and 92%. They indicated that apart from issues about implementing photo enforcement, there were issues regarding privacy amongst the public and judicial concerns about due process. In some jurisdictions, courts dismissed photo enforcement violations because it was not clear that the motorists understood their rights and responsibilities.

ORGANIZATIONAL CULTURE AND BEHAVIOR

Systems Analysis and the Sociotechnical Framework

Yeh and Multer (2007) applied the sociotechnical system framework described earlier in Chapter 1 to the problem of grade crossing incidents. The sociotechnical framework implies a systems analysis, and as such, elements of individual behavior, automation and systems design, the influence of organizations and management, and social and political influences all interact and produce the outcome that we call an incident.

Yeh and Multer discuss motorist decision-making at grade crossings with regard to four elements of the grade crossing system. These include the design of the grade crossing environment, motorist characteristics, the role of organizations and management in grade crossing safety, and the social and political forces that act upon all the other elements.

The grade crossing environment entails the traffic control devices used at grade crossings, including advance warning signs and pavement markings that indicate the proximity of a crossing, passive warning devices such as crossbucks, and active warning devices such as flashing lights and gates. The illumination of the crossing, the available sight distance, and alerting devices on the train (lights, reflective markings, horn) are all part of the grade crossing environment. Although not explicitly mentioned by Yeh and Multer, typical train and highway traffic are also an important element of the grade crossing.

Motorist characteristics are defined by Yeh and Multer as encompassing driving skill and driving style. Driving skill is affected primarily by age and experience. These two factors determine how the driver will cope with distractions and impairment due to drug and alcohol-use and fatigue. Driving style encompasses many aspects of bias as described above in the discussion of signal detection theory and grade crossings: the perceived probability of a train at the crossing, the perceived cost/benefit of compliance (e.g., the waiting time for a train to pass, accident avoidance, etc.), and the motorist's risk tolerance.

Grade crossing safety is highly dependent on the role of numerous organizations and their management. Federal, state, and local agencies need to coordinate with each other and also with the railroad companies. In the railroad industry all of these organizations exercise jurisdiction over some aspect of the grade crossing environment. Railroad companies own and maintain the tracks, the roadway between and around the rails, and the traffic control devices at the grade crossing. State and local agencies and private entities own the roadways at the grade crossing. There are institutional barriers to coordination between these organizations. For instance, railroad engineers and their employers (railroad companies) tend to focus on railroad issues, while highway engineers and their employers (state and local governments) tend to focus on highway issues. Poor communication between organizations contributes to problems such as the failure to properly interconnect highway signals and grade crossing warning devices.

The political and social context concerns regulatory oversight and the development of policies. In addition to the enforcement of traffic regulations, local agencies educate motorists about appropriate driving behavior at grade crossings. State legislatures and courts influence these regulations by their actions in passing laws and judging cases. An example, mentioned above, involved an attempt by the FRA, working with local law agencies, to implement automated photo enforcement. Although this program proved to be highly effective, some judges would not admit photo enforcement evidence because violators may not have been sufficiently informed of their rights and responsibilities. Public reaction to policies and regulations also affect the implementation of safety improvements.

Yeh and Multer illustrated the use of the sociotechnical system analysis with a grade crossing accident at Fox River Grove, Illinois, on October 25, 1995 in which a commuter train struck a school bus at a traffic intersection. The accident resulted in seven fatalities and 25 injuries. The rear of the bus extended onto the tracks, approximately 3 feet into the path of the train. Seven of the 35 students on the bus sustained fatal injuries, 24 students received serious injuries, and the bus driver received minor injuries. The accident report by the National Transportation Safety Board (1996) mentioned all the elements of the sociotechnical system. Yeh and Multer (p. 1553) concluded that "The Fox River Grove grade crossing accident illustrates the value of addressing safety from a systems perspective. While it is always easy to identify system failures in retrospect, the example shows how various layers of the model interact and highlights the impact that organizations can have on safety and the challenges faced in overcoming institutional barriers."

North Carolina Sealed Corridor Project

The North Carolina sealed corridor project, although not a human factors initiative, is another example of organizational action at various agencies and entities to improve safety at grade crossings. This project ran from March 1995 through September 2004 and depended on railroad, FRA, state and local cooperation for its successes.

During that time period high-speed rail passenger service was being encouraged in the United States through Congressional legislation (Bien-Aime, 2009). The proposed Southeast High Speed Rail (SEHSR) Corridor, which connects Washington, DC, to Jacksonville, FL, would use track in North Carolina. To allow train speeds of up to 110 mph, the sealed corridor project intended to improve or close every grade crossing, both public and private,[3] on a route between Charlotte and Raleigh, NC. Along this route there are 216 grade crossings of which 46 are private crossings. In Phase I, II, and III of the project, 208 public crossings were examined and 189 were improved and/or closed (Bien-Aime, 2009). In Phase IV (Bien-Aime et al., 2012) the private crossings were addressed, and 44 of the 46 crossings were improved and/or closed.

Crossings are closed when they are redundant or under-utilized. The process is called consolidation and the FRA has established guidelines for crossing consolidation (Allen, 2009). It is important to note that the guidelines insist on prioritizing jurisdictional authority in the selection of crossings for closure. Such authority can lie in the state or local authorities. Federal resources are only used to provide assistance to the process. Private crossings are often not regulated by state or local authorities, and the landowner and railroad must negotiate the process.

Improvements at grade crossing included traffic channelization, four quadrant gates, four quadrant gates with channelization, video enforcement, long gate arm, and grade separation (Federal Railroad Administration, 1994). Many of these improvements are costly, and Federal funds are available for such projects under various laws. In 2009 dollars (Bien-Aime) the treatments range in value from $85,000 for long gate arms and $4,000,000 for grade separation.

Under Phases I–III, Bien-Aime estimated that 19.7 lives were saved as a result of the project. Under Phase IV, Bien-Aime et al. estimated that 1.5 lives were saved. Considering that the rail route encompassed only 173.3 miles of approximately 140,000 miles of U.S. track, and that this project encompassed 216 of approximately 192,000 public crossings and 46 of 85,000 private crossings, these results are truly impressive.

TRESPASSING PREVENTION

There is a paucity of reliable research on trespassing. A recent review of the literature on trespassing (Topel, 2019) found 102 studies/reports on the topic, of which only 46 were deemed suitable for citation. Generally speaking, the demographics of trespassers is the area of research that has been most thoroughly studied. According to Topel, 70% of U.S. railroad deaths are due to trespassing and suicide. Data on suicides, however, has only recently been collected (as of June 2011) by the FRA's Office of Safety, and that data is not publicly available. Since the Office of Safety's definition of "public" includes the FRA's Human Factors Research Division, suicide data has not been thoroughly examined and analyzed by credentialed, professional scientists. Moreover, a death is only considered a suicide if such a determination is made by coroner, public police office or other public authority (Federal Railroad Administration, 2011). It is known in epidemiological research, that coroners and other public officials are often reluctant to label a death as suicide, which may compromise the usefulness of the suicide data in the FRA database.

CHARACTERISTICS OF TRESPASSING AND TRESPASSERS

Chase et al. (2018) provide a statistical summary of the characteristics of trespassers and trespassing from 2012 to 2014. During that time period, there were 4,430 trespassing incidents of which 2,495 incidents led to a fatality. 782 (31%) of these fatalities were determined to be suicides.

Chase et al. found that the states with the highest number of fatalities [suicides and trespassing without intention to commit suicide (non-intentional trespass)] were California (19.3%), Illinois (7.3%), Texas (6.3%) and New York (5.7%). The most fatalities occurred during the summer (29%), followed by spring (26%), fall (26%) and winter (19%). Suicide fatalities occurred in spring and summer at approximately 30%. Saturday had the highest number of trespasser fatalities with 17%. Suicide fatalities were the highest on Monday (16%). The time periods, 4 pm to 8 pm and 8 pm to 12 am, accounted for at least 20% of incidents individually. The highest percent of suicide fatalities is between 8:00 pm and 12:00 am (22%). 60%–70% of all incidents (both suicides and non-intentional trespassing) involved a freight train, while 30%–40% involved a passenger train. This difference appears to be consistent with the large difference between miles traveled by freight railroads and passenger railroads, although a fatality rate for freight vs. passenger operations would be definitive. More than 80% of both suicide and trespass incidents occurred on the right-of-way, as opposed to a grade crossing.

At least 20% of all incidents occurred in the age group of 15–24 years for each year. Individuals 25–34 years old and 45–54 years old also accounted for a higher percentage of incidents. Individuals from 65 to 74 years old never reached 5% of incidents for any year. The data are not very different between suicides and trespassers.

Walking was the physical act which had the highest percentage of incidents each year (33%–37%). Lying down and standing are next most frequent actions accounting for 25%–28% and 13%–15%, respectively. Walking, lying down and standing account for three-fourths of the incidents overall. By contrast, over 30% of suicides were lying on the tracks, 23 were walking, and 22% were standing.

Additional findings illustrate a number of environmental and individual factors that are associated with each incident, including location (region, state, and right-of-way vs. grade crossing), time (season, month, day of the week, time of day), and characteristics of the individual (age, gender, physical act at the time of the incident).

For suicide fatalities, timing trends emerged that could be attributed to increased opportunities for incidents to occur, for example, during increased passenger train frequency that occurs during typical weekday evening commuting hours. Seasonal timing trends were also found, which follow national suicide seasonal trends which have shown to peak in the spring. In terms of the decedents involved, it was found that age of the decedents is younger than would be expected based on national suicide rates. When compared with incidents that resulted in an injury, it was found that fatalities more often involved a passive action (e.g., lying down, sitting, and standing), while more injuries involved a more active movement (e.g., running, jumping, and walking).

Trespass incidents did not follow the seasonal peak in the spring and summer as with suicide fatalities; rather, trespass incidents peaked in the summer and fall. Additionally, trespass incidents occurred more often on weekends than suicide attempts, particularly when a passenger train was involved. They also followed a slightly different time pattern throughout the day.

When the authors reviewed grade crossing incidents, it allowed them to obtain information that could not be currently gained in other types of incident reports. While most fatalities occurred at higher train speeds and injuries at lower train speeds, this was not always the case. Several incidents resulted in injuries when the train was traveling at higher speeds; likewise, several fatalities occurred when the train was traveling at lower speeds.

Topel (2018) reports that drugs and alcohol were associated with approximately half of all trespasser fatalities, while 37% of suicides had been drinking at the time of death. Moreover, 96% of suicides had mental health disorders.

Martino et al. (2013) conducted 55 psychological autopsies on individuals identified as suicides on railroad right-of-way in the 2007 to 2010 period. The railroad suicides differed from other suicides in a few aspects: they were generally below the age of 50, often lived in close proximity to the railroad tracks, and more often had a mental disorder, substance abuse problem, or both. Martino et al. also found that their sample had limited access to a firearm. The combination of these factors may have determined the choice of suicide on railroad right-of-way to end the individual's life.

AUTOMATION AND SYSTEMS DESIGN

An FRA report by Gabree et al. (2014) outlines a number of systems design countermeasures to prevent trespassing. These include speed restrictions at specific times and locations. Studies which identify when and where trespassing incidents occur can be used to inform this countermeasure (Berman et al., 2013). However, it can be anticipated that railroads will not favor this approach because of operational issues and costs.

Track surveillance is also mentioned by Gabree et al. (2014) as a countermeasure. Efforts are currently underway to automate the detection of trespassers which would reduce the cost of implementing surveillance with human monitors (Federal Railroad Administration, 2022). The success of this countermeasure also depends on a broader participation of organizational elements within the railroad (train crew, train police) and local police or emergency agencies to act to prevent an incident. Consequently, this countermeasure clearly embodies system design, automation and organizational behavior elements.

Gabree et al. (2014) also note that the use of platform screen doors would be effective in deterring trespassing from train platforms. This application is often seen in airports that use people movers. It would be expensive and would only prevent trespassing at stations. The identification of hot spots along railroad rights-of-way could lead to the more extensive use of fencing since access to track is a major contributor to successful rail suicides (Berman et al., 2013). Best practices for the use of fencing can be found in Stanchak et al. (2015).

ORGANIZATIONAL CULTURE AND BEHAVIOR

Federal Railroad Administration (2021) outlines a study that documents an organizational approach to trespass prevention involving stakeholders at the Federal, city and railroad level. The study was an implementation of the CARE model (Community, Analysis, Response, and Evaluation) which FRA and Transport Canada created to develop long-term trespass prevention strategies through collaborative partnerships. The study demonstrated a mostly successful implementation of the CARE model in that an active stakeholder was organized (Community) and provided for pooling of resources to collect information about the trespassing problem and determine its underlying causes (Analysis). This resulted in the identification and implementation of feasible responses to the identified causes of trespassing (Response). The stakeholders recommended strategies on engineering treatments, education outreach activities, and law enforcement actions, including continuing outreach to the homeless population, adding NO TRESPASSING signage, and increasing enforcement patrols. However, the Evaluation phase of the study was not implemented, so the effectiveness of this model is undetermined.

Law enforcement efforts to prevent trespassing necessarily involve multiple actors in federal, state and local agencies and groups. Actions have included education, officer-on-a-train programs, and legal penalties. These efforts appear to reduce risky behaviors along the railroad right of way, but there is limited empirical data on the effectiveness of these actions (Horton and Foderaro, 2016).

NOTES

1 A rate per minute was suggested by the fact that the average freight train is approximately 67 cars long (AAR,1993) and the average train speed through a crossing is 30 mph (Table 55, 1986 Rail-Highway Crossing Accident/Incident and Inventory Bulletin). At an approximate car length of 50 feet, the average train would take approximately 1 min to go through the average crossing.
2 Different grade crossing devices mitigate accident risk, so p(AC) is usually smaller than p(AR) for a given grade crossing device. The ratio, p(AR)/p(AC), provides an easy way to show this. Raslear (1996, Fig. 7) shows that gates have the highest ratio ($\approx$2,000), followed by flashing lights ($\approx$150), special warnings ($\approx$90), and highway signals ($\approx$20). Passive devices have smaller ratios than active devices by an order of magnitude. Finally, grade crossings without any protection (no signage) have a higher probability of accidents than is expected on the basis of train and car traffic.
3 Private crossing include farm crossings, industrial crossings, residential crossings and temporary crossings for construction.

REFERENCES

Allen, L. (2009). *Crossing consolidation guidelines* (Report No. RR 09-12). Washington, DC: U.S. Department of Transportation.

Association of American Railroads. (1993). *Railroad facts.* Washington, DC: Association of American Railroads.

Berman, A., Sundararaman, R., Price, A., Marshall, K., Martino, M., Doucette, A., Chase, S., & Gabree, S. (2013). *Defining characteristics of intentional fatalities on railway rights-of-way in the United States, 2007–2010* (Report No. DOT/FRA/ORD-13/25). Washington, DC: U.S. Department of Transportation.

Bien-Aime, P. (2009). *North Carolina "sealed corridor" Phase I, II, and III assessment* (Report No. 15/3809/17). Washington, DC: U.S. Department of Transportation.

Bien-Aime, P., Carroll, A., & daSilva, M. (2012). *North Carolina "sealed corridor" phase IV assessment: Private crossings* (Report No. DOT/FRA/ORD 12/12). Washington, DC: U.S. Department of Transportation.

Bootsma, R. J., & Oudejans, R. R. D. (1993). Visual information about time-to-collision between two objects. *Journal of Experimental Psychology: Human Perception and Performance, 19*, 1041–1052.

Carroll, A., & Warren, J. (2002). Photo enforcement at highway-rail grade crossings in the United States: July 2000–July 2001. *Transportation Research Record, 1801,* 46–53.

Carroll, A., Multer, J., & Markos, S. H. (1995). *Safety of highway-railroad grade crossings: Use of auxiliary external alerting devices to improve locomotive conspicuity* (Report No. DOT/FRA/ORD-95/13). Washington, DC: U.S. Department of Transportation.

Carroll, A., Multer, J., Williams, D., & Yaffee, M. A. (1999). *Freight car reflectorization* (Report No. DOT/FRA/ORD-98/11). Washington, DC: U.S. Department of Transportation.

Chase, S. G., Hiltunen, D., Scott, H. & Gabree, S. H. (2018). *Characteristics of trespassing incidents in the United States (2012–2014)* (Report No. DOT/FRA/ORD-18/24). Washington, DC: U.S. Department of Transportation.

Church, R. M., & Meck, W. H. (1984). The numerical attribute of stimuli. In H. C. Roitblat, T. G. Bever, & H. S. Terrace (Eds.), *Animal cognition*. Hillsdale, NJ: Erlbaum.

Creelman, C. D. (1962). Human discrimination of auditory duration. *Journal of the Acoustical Society of America, 34*, 582–593.

Federal Highway Administration. (1986). *Railroad-Highway grade crossing handbook* (2nd edn). (Report No. FHWA-TS-86-215). Washington, DC: U.S. Department of Transportation.

Federal Railroad Administration. (2011). *Guide for preparing accident/incident reports* (Report No. RR-22). Washington, DC: U.S. Department of Transportation.

Federal Railroad Administration. (2021). *Trespass prevention research study: Worcester, MA* (Report No. RR 21-08). Washington, DC: U.S. Department of Transportation.

Federal Railroad Administration. (2022). *Railroad trespass detection using deep learning-based computer vision* (Report No. RR 22-04). Washington, DC: U.S. Department of Transportation.

Federal Railroad Administration. (1987). *Rail-highway crossing accident/incident and inventory* bulletin. No. *9, Calender Year 1986*. Washington, DC: U. S. Department of Transportation.

Federal Railroad Administration. (1994). *Rail-highway crossing safety action plan support proposals*. Washington, DC: U.S. Department of Transportation.

Federal Railroad Administration, Office of Safety. (1995). *Florida's train whistle ban*. Washington, DC: U. S. Department of Transportation.

Ford, R., Richards, S., & Hungerford, J. (1998). *Evaluation of retroreflective markings to increase rail car conspicuity: Safety of highway-railroad grade crossings* (Report No. DOT-VNTSC-RR897-PM-98-22). Washington, DC: U.S. Department of Transportation.

FRA. (2005). *Reflectorization of rail freight rolling stock rule*. 49 C.F.R. §224. FRA.

Gabree, S. H., Chase. S., Doucette, A., & Martino, M. (2014). *Countermeasures to mitigate intentional deaths on railroad rights-of-way: Lessons learned and next steps* (Report No. DOT/FRA/ORD-14/36). Washington, DC: U.S. Department of Transportation.

Gibbon, J. (1977). Scalar expectancy theory and Weber's law in animal timing. *Psychological Review, 84*, 279–325.

Gill, M., Multer, J., & Yeh, M. (2007). *Effects of active warning reliability on motorist compliance at highway-railroad grade crossings* (Report No. DOT/FRA/ORD-09/06). Washington, DC: U.S. Department of Transportation.

Hayes, W. L. (1963). *Statistics for psychologists*. New York: Holt, Rinehart and Winston.

Horton, S., & Foderaro, F. (2016). *Law enforcement strategies for preventing rail trespassing* (Report No. DOT/FRA/ORD-16/03). Washington, DC: U.S. Department of Transportation.

Horton, S., Carroll, A., Chaudhary, M., Ngamdung, T., Mozenter, J., & Skinner, D. (2009). *Success factors in the reduction of highway-rail grade crossing incidents from 1994 to 2003* (Report No. DOT/FRA/ORD-09/05). Washington, DC: U.S. Department of

Jenness, J. W., Lerner, N. D., Llaneras, R. E., Singer, J. P., & Huey, R. W. (2006). *Human factors guidelines for intelligent transportation systems at the highway-rail intersection* (Report No. DOT/FRA/ORD-06/26). Washington, DC: U.S. Department of Transportation.

Kaiser, M. K., & Mowafy, L. (1993). Optical specification of time-to-passage: Observers' sensitivity to global tau. *Journal of Experimental Psychology: Human Perception and Performance, 19*, 1028–1040

Keller, A. S., & Rickley, E. J. (1993). *The safety of highway-railroad grade crossings: Study of the acoustic characteristics of railroad horn systems* (Report No. DOT/FRA/ORD-93/25). Washington, DC: U.S. Department of Transportation.

Lerner, N. D., Jenness, J. W., Singer, J. P., Huey, R. W., & Llaneras, R. E. (2007). *Human factors guidelines for intelligent transportation systems at the highway-rail intersection: Technical report* (Report No. DOT/FRA/ORD-07/04). Washington, DC: U.S. Department of Transportation.

Lerner, N., Ratte, D., & Walker, J. (1990). *Driver behavior at rail-highway crossings* (Report No. FHWA-SA-90-008). Washington, DC: Federal Highway Administration, U.S. Department of Transportation.

Luce, R. D. (1986). *Response times: Their role in inferring elementary mental organization.* Oxford: Clarendon Press.

MacMillan, N. A., & Creelman, C. D. (2005). *Detection theory: A user's guide* (2nd edn). Mahwah, NJ: Erlbaum.

Martino, M., Doucette, A., Chase, S., & Gabree, S. (2013). *Defining characteristics of intentional fatalities on railway rights-of-way in the United States, 2007–2010* (Report No. DOT/FRA/ORD-13/25). Washington, DC: U.S. Department of Transportation.

May, R. M. (2004). Uses and abuses of mathematics in biology. *Science, 303,* 790–793.

Meadow, L. (1994). Los Angles metro blue line light rail safety issues. *Transportation Research Record, 1433,* 123–133.

Multer, J., & Rapoza, A. S. (1998). *Field evaluation of a wayside horn at a highway-rail grade crossing* (Report No. DOT/FRA/ORD-98/04). Washington, DC: U.S. Department of Transportation.

Multer, J., Conti, J., & Sheridan, T. (2001). *Recognition of rail car retroreflective patterns for improving nighttime conspicuity* (Report No. DOT/FRA/ORD-00/07). Washington, DC: U.S. Department of Transportation.

National Transportation Safety Board. (1996). *Highway/railroad accident report: collision of northeast illinois regional commuter railroad corporation (METRA) train and transportation joint agreement school district 47/155 school bus at railroad/highway grade crossing in Fox* River Grove, Illinois, on October 25, 1995 (NTSB/HAR-96/02). Washington, D.C.: National Transportation Safety Board.

Ogden, B. D. (2007). *Railroad-highway grade crossing handbook* (Revised 2nd edn 2007). Washington, DC: U.S. Department of Transportation, Federal Highway Administration.

Omar, T., Ngamdung, T., & daSilva, M. (2014). *Driver performance on approach to crossbuck and STOP sign equipped crossings* (Report No. RR 14-20). Washington, DC: U.S. Department of Transportation.

Parasuraman, R., Masalonis, J., & Hancock, P. (2000). Fuzzy signal detection theory: basic postulates and formulas for analyzing human and machine performance. *Human Factors, 42,* 636–659.

Rapoza, A. S., & Raslear, T. G. (2001). Analysis of railroad horn detectability. *Transportation Research Record, 1756,* 57–62.

Rapoza, A. S., Raslear, T. G., & Rickley, E. J. (1999). *Railroad horn systems research* (Report No. DOT/FRA/ORD-99/10). Washington, DC: U.S. Department of Transportation.

Rapoza, A., S., & Fleming, G. (2002). *Determination of a sound level for railroad horn regulatory compliance* (Report No. FRA/RDV-03/28). Washington, DC: U.S. Department of Transportation.

Raslear, T. G. (1988). Temporal discrimination in rats: A parametric examination of memory and switch functions. *Proceedings and Abstracts of the Annual Meeting of the Eastern Psychological Association, 59,* 24.

Raslear, T. G. (1996). Driver behavior at rail-highway grade crossings: A signal detection theory analysis. In A. A. Carroll & J. L. Helser (Eds.), *Safety of highway-railroad grade crossings. research needs workshop: Appendices* (Vol. II, Report No. DOT/FRA/ORD-95/14.2, DOT-VNTSC-FRA-95-12.2, pp. F9–F56). Washington, DC: U.S. Department of Transportation.

Raslear, T. G. (2007). *Intelligent transportation system/positive train control at highway-rail intersections* (Report No. DOT/FRA/RR 07-20). Washington, DC: U.S. Department of Transportation.

Raslear, T. G. (2015). *Using signal detection theory to understand grade crossing warning time and motorist stopping behavior* (Report No. DOT/FRA/ORD-15/02). Washington, DC: U.S. Department of Transportation.

Raslear, T., & Multer, J. (2009). *Effects of active warning reliability on motorist compliance at highway-railroad grade crossings* (Report No. RR09-04). Washington, DC: U.S. Department of Transportation.

Richards, S. H., & Heathington, K. W. (1990). Assessment of warning time needs at railroad-highway grade crossings with active traffic control. *Transportation Research Record: Journal of the Transportation Research Board*, 1254, 72–84.

Stanchak, K., Foderaro, F., & DaSilva, M. (2015). *High-security fencing for rail right-of-way applications: Current use and best practices* (Report No. DOT/FRA/ORD-15/38). Washington, DC: U.S. Department of Transportation.

Sternberg, S. (1966). High-speed scanning in human memory. *Science, 153*, 652–654.

Sussman, D. E., & Raslear, T. G. (2008). Railroad human factors. In D. A. Boehm-Davis(Ed.), *Reviews of human factors and ergonomics* (pp.148–189). Santa Monica, CA: Human Factors and Ergonomics Society.

Swets, J. A. (1996). *Signal detection theory and ROC analysis in psychology and diagnostics: Collected papers.* Mahwah, NJ: Erlbaum.

Topel, K. (2019). *A literature review of rail trespassing and suicide prevention research.* (Transportation Research Circular E-C242). Washington, DC: Transportation Research Board.

Tustin, B. H., Richards, H., McGee, H., & Patterson, R. (1986). *Railroad-highway grade crossing handbook* (2nd edn, Report No. FHWA TS-86-215). Washington, DC: Federal Highway Administration.

United States Government Accountability Office. (2018). *Grade-crosssing safety: DOT should evaluate whether program provides states flexibility to address ongoing challenges* (Report No. GAO-19-80). Washington, DC: U.S. Government Accountability Office.

Yeh, M., & Multer, J. (2008). *Driver behavior at highway-railroad grade crossings: A literature review from 1990–2006* (Report No. DOT/FRA/ORD-08/03). Washington, DC: U.S. Department of Transportation.

Yeh, M., & Multer, J. (2007). Applying a sociotechnical framework for improving safety at highway-railroad grade crossings. In *Proceedings of the human factors and ergonomics society 51st annual meeting* (pp. 1550–1554). Baltimore, MD: Human Factors and Ergonomics Society.

Yeh, M., Multer, J., & Raslear, T. (2009). An application of signal detection theory for understanding driver behavior at highway-rail grade crossings. In *Proceedings of the Human Factors and Ergonomics Society 53rd Annual Meeting.* Santa Monica, CA: Human Factors and Ergonomics Society.

Yeh, M., Multer, J., & Raslear, T. G. (2016). An examination of five grade crossing safety factors on driver decision making. *Journal of Transportation Safety & Security, 8*, 19–36.

Yeh, M., Raslear, T., & Multer, J. (2012) Evaluating the impact of grade crossing safety factors through signal detection theory. In *Proceedings of the Human Factors and Ergonomics Society annual meeting*, October 22–26, 2012, Boston, MA.

Yeh, M., Raslear, T., & Multer, J. (2013). *Understanding driver behavior at grade crossings through signal detection theory* (Report No. DOT/FRA/ORD-13/01). Washington, DC: U.S. Department of Transportation.

Section IV

Railroad Systems and Operations

7 Fatigue and Hours of Service Regulations

BACKGROUND

Fatigue is a complex multidimensional construct characterized by sleepiness and subjective ill-health (Tepas et al., 1997). Several variables play an important role in producing fatigue, including time on task, the duration and quality of sleep, shiftwork and work schedules, time of day when work and sleep occur, and circadian rhythms. Work schedules and shiftwork can cause cumulative sleep loss and disrupt circadian rhythms. All of these variables are known to impair information processing and increase reaction time, which in turn can increase the probability of operator error (National Transportation Safety Board, 1999).

Fatigue has a long history in railroading. At first the simple view was that time on task was the only important variable and that fatigue could be ameliorated by regulating that variable. Hours of service legislation was first promulgated in 1907 (The Hours of Service Act of 1907, as amended in 1976 and 1989, 49 United States Code[1] 21101-21108) to prescribe the length of on-duty and off-duty time for certain railroad employees. The Hours of Service Act provides the legal basis for Federal Railroad Administration (FRA) Hours of Service (HOS) regulations (49 Code of Federal Regulations[2] Part 228). These regulations applied to locomotive train crews, dispatchers, and signalmen (Sussman and Raslear, 2008; Gertler et al., 2013) in both freight and passenger operations. Because HOS rules were based in legislation, it was difficult to modify specific rules even when new scientific evidence supported changes. FRA was the only transportation agency thus constrained and had for many years sought to be relieved of this burden whenever reauthorization legislation was being considered. Finally in 2008, Congress passed the Railroad Safety Improvement Act of 2008 (RSIA, Public Law[3] 110–432, 122 Statutes at Large[4] 4848, 49 U.S.C. § 20101), which allowed FRA the authority to regulate HOS in general, and specifically to determine HOS suitable for passenger and commuter rail service train crews. The Act requires that HOS regulations for passenger and commuter rail service "…shall consider scientific and medical research related to fatigue and fatigue abatement, railroad scheduling and operating practices that improve safety or reduce employee fatigue, a railroad's use of new or novel technology intended to reduce or eliminate human error, the variations in freight and passenger railroad scheduling practices and operating conditions, the variations in duties and operating conditions for employees subject to this chapter, a railroad's required or voluntary use of fatigue management plans covering employees subject to this chapter, and any other relevant factors." (122 Stat. 4864–5). Consequently, RSIA formally acknowledged the role that human factors research plays in the understanding and management of fatigue in the railroad industry.

DOI: 10.1201/9781003500018-11

Prior to the enactment of RSIA, time on task (on-duty time) was the most prominent aspect of work schedules that was regulated. The Hours of Service Act of 1907 was originally dubbed the 16 h rule because it set the maximum amount of time on duty for train service employees at 16 h. The 1907 Act was substantially amended in 1969 and again in 1974. In 1969 the maximum hours on duty was reduced immediately to 14 h, with a further reduction to 12 h in 2 years. In 1974 the amendments established HOS for signal employees (Government Accountability Office, 2011).

These laws were not totally focused on on-duty time (Gertler et al., 2013). As of 1974 and prior to RSIA, a worker who was covered by the act was limited to a maximum of 12 consecutive hours on-duty in a 24-h period. After 12 h on-duty, the worker had to have at least 12 h off duty before returning to work. If the employee worked less than 12 h in a 24-h period (e.g., 11 h 59 min), the required off-duty period was 8 h. The employee was permitted to work a total of 16 non-consecutive hours in a 24-h period provided that the individual had a minimum of 8 h off duty between the two 8-h work periods. Time spent in transportation (deadheading) to duty assignment counted toward on-duty time. Time spent in transportation from duty assignment did not count toward on-duty or off-duty time (limbo time). No provisions limited the number of consecutive workdays or guaranteed rest days.

HOS for dispatching service employees are different from those that applied to locomotive train crews (also called train and engine or T&E workers) and signal employees and have not changed with passage of the RSIA. On-duty time for dispatchers is limited to 9 h, whether consecutive or in the aggregate, in any 24-h period, in operations with two or more shifts. In other words, if the dispatcher has worked for 9 h, he or she must have 15 h of rest.

As early as 1999, it was recognized by NTSB that railroad HOS rules allowed employees to work an extraordinary number of hours per month. "A commercial pilot may fly up to 100 h per month; a truck driver may be on duty up to about 260 h per month; licensed individuals on an oceangoing vessel or coastwise vessel of not more than 100 gross tons (GT) may operate up to 360 h per month when at sea; and locomotive engineers may operate a train up to 432 h per month". (NTSB, 1999, p. 2). In other words, railroad workers in safety critical positions could work as many as 14.4 h per day, which ignores the possibility (as noted above) of working 16 h in 24 h on split shifts (popular with commuter railroads). Note that in 1999 there was scant research as to how many hours railroad employees actually worked or the schedules under which they worked. The major exception to this was a government report by J. K. Pollard (1996) on sleep and work patterns of freight locomotive engineers and two U. S. Government Accounting Office (GAO) reports (1992, 1993. A critical review of the GAO reports can be found in Raslear, 2019). However, it was clear that HOS regulations that prescribed on-duty time could result in an excessive number of on-duty hours that might result in employee fatigue.

Consequently, in the 1990s the FRA, under the leadership of Administrator Joline Molitoris, recognized that a new approach was needed (Coplen and Sussman, 2000). The new approach emphasized cooperative partnerships with railroad management, labor unions, other federal agencies, state governments, and Canadian federal agencies. Prior to this change in approach, as is obvious from the previous discussion of laws and regulations governing railroad safety, railroads were cited for violations of regulations and civil penalties (fines) were issued.

The premier partnership that was initiated under Administrator Molitoris was the North American Rail Alertness Partnership (NARAP). As outlined in its charter, the purpose of North American Rail Alertness Partnership (Federal Railroad Administration, 1998) was to support industry-wide initiatives through the coordination, facilitation, and communication of efforts to reduce fatigue and promote safety in rail operations and to promote the safety of the industry's employees by developing effective countermeasures, based upon analytical and/or scientific data.

NARAP was, in part, an effort by FRA to capture and dominate the field of fatigue in the railroad industry. This was precipitated by the formation of the Work/Rest Review Task Force[5] in the early 1990s by the Association of American Railroads (AAR[6]). The mission statement of the Task Force was to "Create an environment which enhances an employee's ability to better manage their time off duty, thereby insuring that employees will report for work in a condition to perform their duties more safely and productively." (AAR, 1993). The FRA was not included in the membership of the Task Force, and as is clear from the mission statement, fatigue was framed as the fault and responsibility of the employees. At this time, neither the railroads nor the labor unions wished to see changes in the hours of service because both benefited from the *status quo*.[7] If duty hours were restricted and rest time was expanded, the railroads would be forced to hire and train more employees, and the union members would lose pay. Thus, the cooperation between rail labor and management and the exclusion of the regulator from the "research" project. The product of the Task Force will be discussed in detail later in this chapter.

Although NARAP reinserted FRA into industry discussions about fatigue in the industry, scant progress was made in the conduct of actual research. At least 13 meetings were held from 1996 to 2000 at various locations in the U.S. and Canada. The FRA repeatedly offered funds for research projects, but the offers were always rebuffed. At one meeting, the FRA team was told that funds would only be accepted if the AAR controlled the project (including selection of the contractor, experimental design, statistical analysis, etc.). The FRA team was told that "He who pays the piper, calls the tune", implying that the regulator could not be trusted to conduct an impartial research project. Furthermore, no Class I Railroad would agree to provide access to employees so that work schedule data could be collected.

RSIA changed the statutory limits on work hours for T&E workers in freight service (freight T&E workers), effective July 16, 2009, and kept T&E workers in intercity or commuter rail passenger service (passenger T&E workers) temporarily subject to the pre-RSIA statutory limits. As noted above, RSIA gave FRA the authority to prescribe regulations for passenger T&E workers that differ from the amended statutory requirements applicable to T&E workers in freight service.

With respect to freight T&E workers, RSIA establishes limits per calendar month on service performed for a railroad and on time in or awaiting deadhead to final release. RSIA increases the quantity of the minimum off-duty period after being on duty for 12 h in broken service from 8 h of rest to 10 h of rest. It prohibits railroads' communication with workers during certain minimum statutory rest periods and mandates 48 h time off duty for workers after initiating an on-duty period on 6 consecutive days, or 72 h after initiating an on-duty period on 7 consecutive days (49 U.S.C. § 21103).

New, non-statutory, regulations on HOS for passenger T&E workers went into effect on October 15, 2011 (49 CFR 228; Federal Register Number 2011–20290). The new regulations leave unchanged the limitation of 12 consecutive hour of time on duty in a 24-h period and the mandatory off-duty periods, but limit the number of days of consecutive work for those who work overnight assignments. Some commuter and passenger service includes overnight work, and the rule defines a "Type 2" assignment as one that includes time on duty between 8 p.m. and 4 a.m. If the employee's work schedule includes at least one Type 2 assignment, there is a new limit on the number of consecutive days of work. Also, the work schedules of passenger T&E employees of railroads operating commuter and passenger service must be analyzed using an FRA-approved model of human performance and fatigue to identify schedules with an unacceptable level of fatigue exposure. Work schedules which exceed the fatigue threshold more than 20% of the work time must be identified, and the railroad must propose fatigue mitigation actions for those schedules, subject to FRA approval. Much of the passenger T&E rules were a direct result of the research conducted by the FRA's Human Factors Research Division (and its predecessors), as mandated by the section of RSIA quoted on p. 99 above.

The research described in this report below was conducted prior to the changes in the HOS Law described above.

FATIGUE RESEARCH IN THE RAILROAD INDUSTRY

EARLY EFFORTS

The Government Accounting Office (GAO) Reports

Fatigue has been an issue for a long time in railroading, as noted above. At FRA, complaints from labor unions about working conditions, including fatigue and violations of HOS rules, were common. Also there were catastrophic accidents, such as the Hinton train collision in 1986 in which 23 people were killed and more than $30 million in damage was caused. Analysis (Smiley, 1990) of the Hinton collision indicated that human error was the cause of the collision and that the major factor was poor work-rest schedules (i.e., fatigue). In 1992 and 1993, the GAO issued reports on locomotive engineer work schedules and human factor accidents at the request of the Committee on Energy and Commerce, House of Representatives. The 1992 report attempted to answer the following questions:

> Are railroads complying with the … requirements of the Hours of Service Act?
> Can safety be improved by amending the act to reduce the maximum number of hours (e.g., from 12 to 10) than an engineer is allowed to work?
> Do work schedule factors other than the maximum number of hours allowed by the act affect safety? (GAO, 1992, p. 1)

The Committee further requested that GAO address:

1. how the timing and frequency of human-factor-caused accidents that occur on yard tracks differ from those on mainline tracks;

2. how accident rates differ depending on the type of shift, the time of day, the hour of the shift, and the shift's start time variability;
3. how work schedule characteristics differ between shifts worked by engineers who had accidents and shifts worked by the general population of all engineers; and
4. how the mechanism for notifying train crews to come to work, or "crew-calling system," affects the ability of engineers to predict their next work shift. (GAO, 1993, pp. 1–2)

The questions posed by the Committee were all excellent questions which, unfortunately, the GAO did not entirely adequately answer. Time pressures, lack of personnel with requisite skills in statistical analysis, and the lack of tools to assess employee fatigue based on work-rest history, probably all contributed to inadequacies in GAO's reports.

The GAO (1992) data are based on locomotive engineer work schedules from Atchison, Topeka & Santa Fe Railroad, Kansas City Southern Railroad, Southern Pacific Railroad and Burlington Northern Railroad. These are all Class I railroads, but they all operate in the western U.S. which makes the data difficult to generalize to railroads that operate in the eastern U.S. or are not Class I railroads.[8] GAO sampled 124 work schedules for the study.

With regard to the first question for GAO 1992, the report estimated that 99.4% of work periods of 12 or more hours were followed by at least 10 h off-duty.

With regard to the second question, GAO 1992 reported that the percent of human-factor accidents varied with the hours on duty (GAO, 1992, Figure 1). Accident percent peaked at 3 h of duty time (12.6%), declined to 2.5% at 10 h, and to 0.1% at 12 h. The report concluded that restricting on-duty time to 10 h would have little effect on accidents. This analysis is misleading and inaccurate (see Raslear, 2019) because the frequency with which shifts are worked is not constant for different shift durations, as can be seen in Figure 2 of GAO/RCED-92-133. A more accurate picture of accident occurrence as a function of shift duration would be obtained by plotting accidents/shift as a function of shift duration. This would normalize the accident data with regard to the frequency of shifts as a function of shift duration (see chapter on exposure). This inadequacy is acknowledged (but not remedied) in GAO/RCED-93-160BR: "Accident rates (accidents per 100,000 engineer work hours) provide a better understanding of the likelihood of accidents than a straight count of the number of accidents that occur." (p. 13).

Figure 7.1 shows the Accidents per Shift as a function of Shift Duration (h) for the GAO 1992 data. The normalized accident data indicates that the accident rate is highest in the first 2 h of work and declines rapidly thereafter. The data resembles an exponential decay function, and a least squares fit of the data to that model indicates that an exponential model is an adequate description of the data ($r = -0.9723$, $df = 5$, $t = -9.5$, $p < 0.01$). An exponential form to the accident/shift data suggests that accidents are random events which are independent of time (Luce, 1986, p. 15). In other words, accidents are independent of shift duration according to this data.

The analysis in Figure 7.1, however, is also inaccurate in that it under-represents short work durations. For instance, an engineer who works 12 h is only counted in the 10.1–12 h bin, but has obviously worked during that shift for 2 h, 4 h, etc., until the

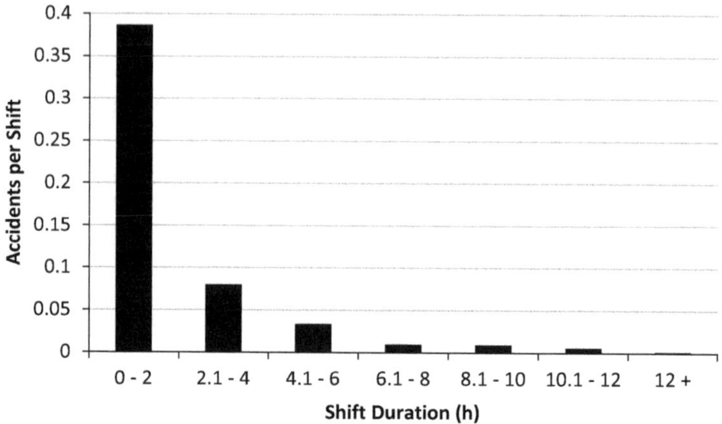

FIGURE 7.1   Accidents per shift as a function of shift duration (hours).

shift has ended. Since an accident could occur at any time during a 12-h shift, the total shift-hours for each bin should include counts of longer duration shifts to accurately represent the number of occasions on which an accident could have occurred (shift-hours at risk). Following this reasoning, the number of shift-hours in the 0–2 h bin is 100,000 (the total number of shifts), in the 2.1–4 h bin it is 98990 (=100000–1010), and in the 4.1–6 h bin it is 92630 (=98990–6360), etc. In statistics (Evans et al., 2000; Luce, 1986) this is known as the survivor function $(1 - F(t))$, and it is the compliment of the distribution function $(F(t))$. If one divides the number of accidents per shift duration (this is the density function for accidents, $f(t)$) by the survivor function, the result is the hazard function $(\lambda(t))$. A hazard function is the tendency for an event to occur over time (Luce, 1986). Figure 7.2 shows the hazard function for accidents (accidents per shift-hour).

From Figure 7.2 it is clear that the true accident rate is fairly independent of duration of work. The only odd point occurs for the bin 12+ which shows a much lower rate. This, however, may be due to the small number of accidents (4.154, see Raslear,

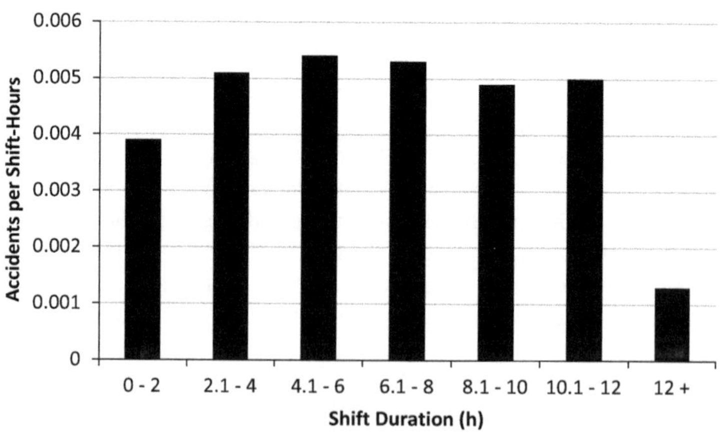

FIGURE 7.2   Accidents per shift-hour as a function of shift duration (hours).

2019) available to compute the accident rate. A regression analysis, moreover, suggests that there is no trend in accident rate as a function of work duration ($r = -0.45$, $df = 5$, $t = -1.13$, $p > 0.05$). Since the hazard function is statistically flat, this suggests that accident risk is independent of the duration of a shift or the time worked on a shift.

With regard to the third question, GAO 1992 found that half of the engineers' schedules they examined varied in start times by at least 2 h, and that 30% of start times varied by at least 6 h. GAO 1992 speculated that start time variability could disrupt circadian rhythms and contribute to engineer fatigue.

GAO 1993 (GAO/RCED-93-160BR) had a more extensive list of questions to answer from Congress (see above). Although, GAO 1993 added two eastern railroads (Conrail and CSX) to the data from GAO 1992 and shift data from nearly 8,000 more engineers, these two railroads only added 75 additional accidents for analysis. Consequently, several problems arise in analyzing GAO 1993 data alone or in conjunction with GAO 1992 data. These are documented in Raslear (2019) and include the sample size in GAO 1993, the lack of random sampling in GAO 1993, the lack of information on statistical methods and results in GAO 1993, the use of different samples for accidents and shifts in GAO 1992, the lack of a comparison (control) group in GAO 1992, and the arbitrary exclusion of Union Pacific data in GAO 1992.

The Pollard Reports

John K. Pollard (known to colleagues as JK) conducted two seminal studies concerning fatigue and work schedules in locomotive engineers (Pollard, 1991, 1996) as a staff member of the Volpe National Transportation Systems Center (Volpe), sponsored by FRA's Office of Policy. Although neither report is definitive, Pollard's methods demonstrated how the practice of good science could illuminate contentious human factor issues in the railroad industry.

Pollard (1991) examined the crew calling and scheduling systems in place in the seven Class I railroads. At the time, there was no research and information in the scientific literature about these topics. Fred Gamst (1980) had published an industrial ethnography about the railroad industry, but that book did not discuss crew calling and scheduling systems. Because of the lack of previous research, Pollard utilized what are today called qualitative methods to obtain basic information. He used two methods, interviews and focus groups (Berg, 1995; Kerlinger, 1973; Madill, 2012; Morgan, 1988).

From interviews with operating managers, dispatchers, crew callers, and union officials, Pollard was able to describe how locomotive crews were scheduled at that time. The mechanics of scheduling, the unpredictability of schedules, and timing of scheduling decisions were documented. The problems of fatigue and stress, and how the scheduling systems in use contributed to these issues, were discussed. Focus groups with locomotive engineers revealed that the scheduling system led to uncertainty in the start time of the next job, long working hours, commutes and waiting times. The engineers also described the poor condition of locomotives and sleeping accommodations at terminals. These issues, and others, contributed to the fatigue and stress experienced by the engineers. The GAO 1993 report relied heavily on Pollard's description of crew calling and scheduling in the Class I railroads.

Pollard (1996) pioneered the use of work/rest diaries and subjective alertness scales in the railroad industry. Based on his 1991 research report, Pollard sought to obtain quantitative information concerning work schedules, working conditions, fatigue and stress. He collected a convenience sample of 198 locomotive engineers (out of 800 diaries distributed) from whom he obtained a 14-d work/rest diary, demographic information, information on commute time, type of job (yard, regular road pool, etc.), alertness at various times during work, activity type during the day (work, personal, etc.), quality of sleep, and information about calls to duty. Based on this information, Pollard asked the following questions:

How do engineers divide their time among such activities as:

working
sleeping
commuting
all other personal time?
How alert do they feel on the job?
What is the quality of their sleep:
at home
away from home?
How well can they predict when their next jobs will start?
How do job-start and job-end times affect sleep?
How many miles do they travel?
How does having an assistant engineer affect self-rated alertness? ...
How does alertness vary with:
time of day
employer
workload?
How does average sleep vary with:
age
place of rest
employer
workload? (Pollard, 1996, p. 3).

We will focus on the results obtained by Pollard as regards time on- and off-duty, sleep and alertness. Those interested in the other topics are encouraged to read the original report.

Pollard found that the engineers worked 37.8 h per week on the average (which agrees roughly with the GAO 1993 findings of 35.9 h), but that some engineer groups worked less than 30 h per week (Railroad A extra board) and others worked more than 50 h per week (Railroad F road assignment). By comparison, GAO 1993 found that 25% of the 6,996 engineers worked less than 23.2 h and 5% worked 72.65 h or more. 60.1% (119) of the engineers worked in regular pools. 9.6% (19) worked road assignments. 23.7% (47) worked the extra board, and 8.6% (17) worked exclusively in yard or local service.

The Pollard study asked the engineers about the amount of sleep they got at home and away from home. They averaged 7.13 h total daily sleep. By comparison, the U. S. population gets 7.5 h total daily sleep. Sleep periods[9] at home averaged 6.58 h, and sleep away from home averaged 6.08 h. Type of work assignment affected total daily sleep: 6.78 h for yard engineers, 6.95 h for extra board engineers and 7.2 h for regular pool engineers. Time of day when work starts and ends also affects total daily sleep. Engineers with job starts between 0800 and 1500 h got the most sleep (between 7 and 8 h). Engineers with job starts in the night or early morning got much less sleep. Pollard reports that "Job starts between 2200 and 0300 are associated with average total sleep of only about 5 h." (1996, p. 16).

Engineers were asked to report their alertness in their survey books every 2 h. A four point Likert scale was used in which 1 indicated being fully alert, 2 indicated being moderately alert, 3 indicated being drowsy, and 4 indicated fighting sleep. Figure 7.3 shows the self-rated alertness of the engineers as a function of time of day. The data has been double plotted to emphasize its circadian rhythmicity. It is pretty clear that time of day is a significant predictor of alertness in this data. The lowest alertness occurs in the morning hours between 0400 and 0600. The highest levels of alertness occurs between 1600 and 1900. The GAO 1993 report accident rate data were inconclusive in showing a circadian rhythm because of the small number of accidents, and the GAO 1992 report accident data showed the highest accident rate between 0200 and 0600. Both reports had insufficient data to adequately document a circadian rhythm convincingly. It is clear today that alertness, sleep and accidents are critically related.

Pollard in his seminal report was the first to link sleep and alertness to time of day.

Pollard also found a small but reliable correlation (0.1519, $p < 0.05$) between alertness and average daily work hours. A correlation of this magnitude indicates that knowledge of the duration of an engineer's work day only accounts for 2.3% of the variance in alertness. In other words, work day hours were a poor predictor of alertness. Again, since alertness is, today, known to be a factor in accidents, this result is not inconsistent with the data from the GAO studies on time on duty – both show that duration of work alone is not a major factor in fatigue and accidents.

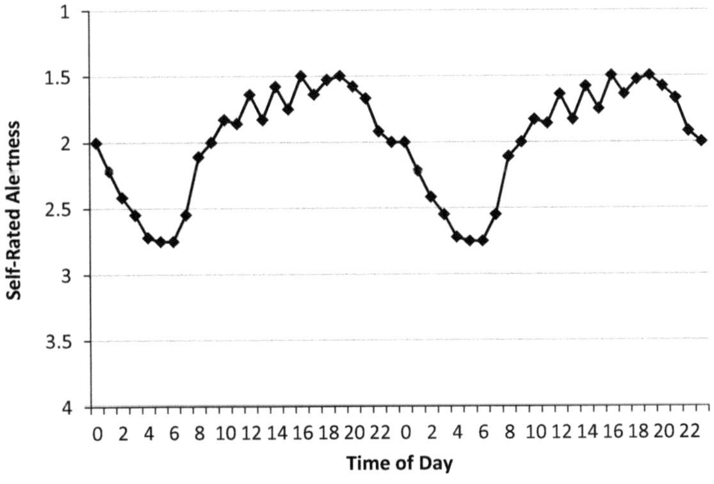

FIGURE 7.3   Self-rated alertness as a function of time of Day. The data is double plotted to emphasize the circadian rhythm in the data. After (Pollard, 1996).

The AAR sponsored Work/Rest Review Task Force Research

On September 9, 1993 a meeting was held in FRA headquarters with representatives of the AAR, the Brotherhood of Locomotive Engineers (BLE), the United Transportation Union (UTU), the FRA (Office of Safety, Office of Research and Development, Office of Policy), and Volpe. The purpose of the meeting was to discuss FRA and AAR efforts to understand the factors that contribute to fatigue in locomotive engineers (AAR, 1993; Raslear, 1993). The FRA discussed work/rest diary research being conducted by Volpe (later reported in Pollard, 1996) and research on fatigue and train handling being conducted in the locomotive simulator (discussed below) at the Illinois Institute of Technology Research Institute (IITRI). The AAR discussed the plans for the research to be conducted by the Work/Rest Review Task Force.

According to the AAR (1993), "The task force members have agreed on a disciplined statistical study that will "benchmark" existing crew work schedules throughout the industry and then compare this benchmark to the schedules of crews who have experienced accidents, injuries or a rule violation.

The work schedules of a randomly-selected group of engineers from a randomly— selected group of terminals on several railroads are being gathered and examined. Next, the schedules of engineers who have experienced accidents during the same time period covered by the benchmarking analysis will be examined to determine any measureable differences."

Raslear (1993) noted that AAR is "...in process of collecting data from 1,200 subjects on sleep/work etc. Sounds like no research plan is in place." In a letter dated November 4, 1994, Edwin Harper, the president of AAR (AAR, 1994) provides a status report on the Work/Rest Review Task Force. The letter states that: "The Task Force has neither found any overall statistical correlation between accidents and incidents and crew time on duty or time of day, nor has it found any statistical correlation between incidents and types of work schedules other than those noted. ...

Between midnight and 6 a.m., the potential for accidents increases when a train and engine crew has been on duty for greater than 9 h.

The potential for an incident increases when a train or engine service employee has worked:

- five consecutive permissible shifts with an average shift length greater than 10 h
- more than six consecutive permissible starts in a 7 day period."

According to the attachment to the letter, the Task Force had collected 70,000 shifts (five consecutive shifts for each subject, indicating that more than 1,200 crew members' work histories were collected). Work schedules were evaluated by start time variability, shift length, and shift density (number of shifts). The Task Force suggested a graphical method of combining and analyzing these three "shift factors". The method assumes that each of the factors has equal weight. Table 7.1 shows the Combined Shift Factor (CSF) for Regular Jobs (8 h work per day, 5 days per week, no variability in start time), Short Pool jobs (10 h per day, 6 days per week, ±4 h start time variability), and Long Pool jobs (12 h per day, 3 days per week, ±5 h start time variability), respectively. The CSF is calculated as the area of the triangle in each graphic,

TABLE 7.1

Combined Shift Factor (CSF) for Regular, Short Pool and Long Pool Jobs Based on Shifts Worked, Shift Start Time Variability and Hours on Duty

	Regular Job	Short Pool	Long Pool
Shifts worked (A)	5	6	3
Variability (B)	0	4	5
Hours on duty (C)	8	10	12
CSF ($0.5 \times (A \times (B + C))$)	20	42	25.5

Source: Data From (AAR, 1994).

$CSF = 0.5 \times (A \times (B + C))$, where A is the number of shifts worked, B is the variability of shift start times and C is the number of hours on duty. The height of the triangle is A and the base of the triangle is the sum B + C.

There is no indication in the AAR materials as to how it was determined to break the 70,000 shifts into the three categories described in Table 7.1, how Short and Long Pool jobs are defined, or whether there are other types of jobs that the employees work. Also note, that despite the statement that accidents have a greater potential to happen between midnight and 6 a.m., there is no time of day variable included in the CSF. It is also not known how many work histories are included in each of the three types of jobs.

Based on this analysis, the AAR letter suggests eight "strategies" to counter the potential for accidents and incidents:

Bulletins issued to crews
Crew members notified by crew dispatcher at call for duty
Job briefing prior of going on duty
Train dispatcher alerting crews when they are approaching the ninth hour
Adapt automated crew dispatching systems to "flag" employees who are approaching the thresholds
Provide specific employee education for those who are approaching the thresholds
Include these issues in safety and job briefings

None of these suggestions deal explicitly with the main problem – changing the schedules that train crews work so as to reduce fatigue.

The complete results of the Work/Rest Task Force were never shared with the FRA and were never made public. The research design, selection of subjects, statistical analyses, and other critical details of the project remain unknown. Consequently, the scientific accuracy of the conclusions drawn from this project cannot be evaluated and confirmed.

FATIGUE RESEARCH SPONSORED BY THE HUMAN FACTORS RESEARCH GROUP

Human factors research was initially subsumed under the Equipment and Operating Practices Division of FRA's Office of Research and Development, led by Claire Orth (Chief). Over time the Operating Practices segment of the Division became the Human Factors Research Branch (Raslear as Staff Director), and finally the Human Factors Research Division (Raslear as Chief). During this 20+ years period of time, the principal staff members were Garold Thomas, Thomas Raslear, and Michael Coplen. When Garold Thomas retired, Coplen was hired (he had worked as a contractor and Volpe staff member previously). Subsequently, Mike Jones and Starr Kidda were also hired to work in the Division. With the exception of some early work, the bulk of the fatigue research was planned and executed by Coplen and Raslear.

There was a great deal of attention at this time (1990s) being paid to the topic of operator fatigue. The Congress was concerned because of high profile accidents such as the Hinton Train Disaster (Smiley, 1990), the pressure from labor unions to modify HOS laws and regulations, the announcement by NTSB that fatigue was on its "Most Wanted List" (Raslear et al., 2013) and the GAO reports in 1992 and 1993. The FRA and the Department of Transportation (DOT) were concerned because of Congress' attention to the fatigue issue and the AAR's formation of the Work/Rest Review Task Force. Amongst the Human Factors (HF) staff, all of this attention meant that the HF staff would eventually be asked to provide research to support a change in legislation and/or regulations regarding HOS. This feeling was shared DOT-wide amongst HF staff in the DOT agencies that dealt with operator performance. Consequently, the DOT Human Factors Coordinating Committee[10] (HFCC) commissioned a contractor to canvass HF staff at all concerned DOT agencies to determine the primary policy issues in fatigue management for each agency and the DOT in general.

Dr. Susan Labin and her research team from the Institute for Survey Research, Temple University (ISR) gathered stakeholder input (industry as well as government) for developing the Cross Modal Fatigue Management Model (Labin, 2005). The Cross Modal Model was intended to be used for long-range strategic program and evaluation planning, knowledge development, and consensus building with stakeholder groups. The ultimate purpose of the model was to support the achievement of long-term goals and outcomes; viz., to reduce fatigue and fatigue related health and performance problems in transportation.

The Cross Modal Model was based on logic models developed for each agency. Logic models visually display implied causal linkages that underlie the purpose and intended results of programs. The Rail Fatigue Management Program and Evaluation Model has three major elements common to logic models: 1. needs and policy issues; 2. programs; 3. results. The macro model (big picture) for rail is shown in Table 7.2.

Table 7.2 and its accompanying report were never published, but the logic model served as a roadmap for FRA's research plan on fatigue management. The logic model contains the same elements as the Human Factors R&D program structure described in Chapter 1. The difference is that Table 7.2 is more detailed and contains the indicators of policy needs and the elements of a strategic plan to resolve them.

For instance, the first column of Table 7.2 identifies, in the first row, that there is no common definition of fatigue, that fatigue is under reported in the FRA accident/

TABLE 7.2

Rail Fatigue Management and Evaluation Program Logic Model

Indicators of Policy Needs	Needs/Policy Issues	Program Type	Policy Audience and Target Populations	Program Outcomes
No common definitions. Under reported as accident cause. Lack of data on fatigue practices.	Define and measure fatigue and countermeasure results. Compile and synthesize existing data.	Fatigue indicators and measures. Blameless accident reporting. Measure program outcomes (costs and benefits). Program outcomes database.	Policy and Program Audience:	Measures: Health, fatigue, stress Safety, accidents, incidents, deaths, injuries Performance, human error
Lack of schedule predictability. Sleep deprivation. Outdated hours of service. Collective bargaining stalemate.	Scheduling reform.	Participatory processes with worker input. Legislation updating hours of service. Programs to improve predictability. Validated fatigue modeling and scheduling. Positive Train Control.	RR industry Management Fatigue managers Schedulers Unions Government FRA -Office of R&D	Quality of life, social, family time Divorce rate Efficiency Mobility Security Costs (e.g., health care)
RR culture resistant to change. Demographic changes in industry. Age and experience predict accidents.	Understand and influence the organizational work culture.	Fatigue education programs. Scheduling that considers age and practices regarding fatigue. Successful non-prescriptive program example.	-Office of Safety -Office of Chief Counsel Target Populations Industry Shift workers	Benefits (e.g., improved worker health, retirement, decreased worker absences) Data Sources:
Inability to assess one's own level of fatigue.	Use technology to provide a "safety net" against human error.	Fatigue monitoring equipment. Measure effects of rest on performance. Control environmental influences (e.g., temperature, noise).		Physiological (e.g. actigraph) Self report (diary) Administrative data

Source: After (Labin, 2005).

incident database, and that there is a lack of data on industry fatigue practices. The need to define and measure fatigue and to compile and synthesize data sources on fatigue follows in the second column of row 1. Programs to meet these needs are suggested in the third column: conduct research to measure fatigue; determine the cost of fatigue to industry and the benefits of fatigue management; devise a program to allow employees to report accidents without repercussions from rail management or government. All of these programs were ultimately implemented, as will be described below. The two remaining two columns identify target audiences and possible program outcomes.

The HFCC defined fatigue for the DOT as follows: "Fatigue produces physical and mental decrements in alertness that can reduce job performance effectiveness, decrease the ability to safely operate vehicles, and increase the risk of human error resulting in fatalities and injuries." (HFCC, 2003).

Railway Worker Fatigue and Performance

The Simulator Study

Much of the FRA's research on fatigue has focused, naturally, on the individual railway worker. However, because of the interaction of the basic elements of the sociotechnical system described in Chapter 1, it is difficult to isolate factors that affect fatigue in the individual worker from system design issues and organizational culture. This can be clearly seen in the study sponsored by FRA to demonstrate that HOS regulations, even if strictly followed, would result in cumulative sleep debt, decreased alertness, and decreased performance. To avoid exposing locomotive engineers to potential danger, the study was conducted in the Research and Locomotive Evaluator/Simulator (RALES).

George Kuehn designed and conducted the RALES simulator study (Thomas et al.,1997). Kuehn devised two week-long work schedules that were intended to produce sleep deprivation and circadian disruption in locomotive engineers. In both schedules the start times of work rotated backwards in time (i.e., shifts started earlier each day). In the "slow" schedule, work starts rotated backwards by 2 h each day. In the "fast" schedule, work started 4 h earlier each day. In both schedules there was 10 h of work, but off-duty time was different. On the "slow" schedule, engineers averaged 6.1 h of sleep, while engineers on the "fast" schedule average 4.6 h of sleep. In both cases, this amount of sleep was well below the average for most people. Sleep characteristics were determined by the use of actigraph watches. The RALES simulator was used to assess the performance of the locomotive engineers in operating a train across a 190 mile route each day (a run) for four days.[11] The train was a coal train consisting of four locomotives and 88 cars.

Subjective alertness was assessed with the Global Vigor scale (Monk, 1989) ten times during each run over the 190 mile route. Time of day influences on alertness (predicted alertness) were modeled with a simple cosine function with a 24-h period. An offset was added to the cosine model to keep the output a positive number. The offset was an exponential function of the sleep debt incurred by the engineers, with a range between zero and one. The model accumulated 1 h of sleep debt for each hour awake and repaid sleep debt 1 h for each hour asleep. Predicted alertness was 100 times the offset plus the cosine function. There was a reliable correlation between subjective alertness and predicted alertness ($r = 0.83$, $p < 0.01$). There were no reliable differences in subjective alertness between the "fast" and "slow" groups, although the effects of runs and the interaction of group and runs were reliable.

The performance measures included missed horns (not sounding the horn at grade crossings), latency to reset the alerter[12] and fuel consumption. Predicted alertness showed reliable correlations with missed horns and fuel consumption. Overall, the study demonstrated that work schedules that conformed with HOS regulations could produce sleep debt in locomotive engineers resulting in fatigue, loss of alertness and decrements in safety critical and economically critical performance. For instance, fuel consumption was increased by 100 gallons for every hour of lost sleep (see Raslear, 2014a, slide 25). The HOS regulations, their implementation at the railroads and their enforcement by the FRA constitute the system design for work conditions for locomotive engineers and the organizational culture which endorses these conditions as "the way we do business in the railroad industry".

Measuring Fatigue

Following the RALES simulator study it became apparent that a method to measure fatigue was needed. The model employed in the RALES study was adequate for the purpose it served, but it was simplistic and depended on actigraph data to determine sleep durations. If a general understanding of fatigue in the railroad industry was to be developed, data from a large random sample of engineers would be needed, and actigraphs would be too expensive to do an economically feasible study. Moreover, it was deemed important to demonstrate that any fatigue model (measurement system) actually measured fatigue (i.e., demonstrated validity[13]) and that the predictions from the model could be related to the risks of meaningful failures of human performance (i.e., was calibrated). Criteria and procedures for determining that a biomathematical model of fatigue was valid and calibrated were formally documented in Raslear (2013b). It is worth noting that sleep science had, prior to 2013, already established informal criteria for evaluating the validity of fatigue models. A 2002 workshop cosponsored by the U.S. Department of Transportation and U.S Department of Defense on fatigue and performance modeling (Neri, 2004) invited presentations of biomathematical models that demonstrated sensitivity to circadian, sleep deprivation, sleep recovery, and sleep inertia effects on one or more well-known behavioral or performance-based indicators of fatigue, including reaction time, cognitive throughput, lapses, alertness, and tendency to fall asleep (Balkin et al., 2002; Balkin et al., 2000; Bonnet, 1997; Carskadon and Dement, 1977; Dinges et al., 1985; Dinges and Powell, 1985, 1989; Folkard and Akerstedt, 1987, 1992; Froberg, 1977; Harrison and Horne, 1996; Jewett, 1997; Jewett and Kronauer, 1999; Lumley et al., 1986; Mitler et al., 1982; Monk, 1991; Monk and Embry, 1981; Richardson et al., 1978; Thorne et al., 1983; Wesensten et al., 1999). FRA considered such demonstrated sensitivity to be a prerequisite for a biomathematical fatigue model to be considered as a candidate for measuring fatigue in railroad operations.

The FRA funded a project (Hursh et al., 2006, 2008) to use the work histories of locomotive crews involved in accidents to validate and calibrate the Sleep, Activity, Fatigue, and Task Effectiveness (SAFTE) biomathematical fatigue model. The U.S. Army and U.S. Air Force developed SAFTE to predict potential fatigue caused by work schedules (Hursh et al., 2004). Subsequently, a fatigue assessment tool (Hursh et al., 2004) called the Fatigue Avoidance Scheduling Tool (FAST) was developed from SAFTE. The FAST tool uses work schedule information to estimate the amount of sleep and cognitive effectiveness.[14] In FAST, cognitive effectiveness is interpreted as the inverse of fatigue. Note that the FAST tool estimates amount of sleep, eliminating the need for actigraphs. SAFTE was one of six biomathematical fatigue models that were invited to the fatigue and performance modeling workshop (Neri, 2004).

The purpose of the project was to demonstrate that the model could predict an increased likelihood of human factors accidents relative to nonhuman factors accidents under conditions of fatigue. It was reasoned that FAST would predict higher levels of fatigue when there was a greater likelihood of a human factors accident. Fatigue levels in FAST are based on opportunities to sleep and the time of day of an accident. Nonhuman factors accidents should have no relationship (or a weaker relationship) to fatigue levels. Cognitive effectiveness (fatigue) was determined from 30-day work histories of locomotive crews involved in 400 human factors and 1,000 nonhuman factors accidents prior to the accident. In SAFTE and FAST, cognitive effectiveness is interpreted as the inverse of fatigue.

The effectiveness of the crew at the time of the accident and the overall percentage of time typically spent at work at various levels of effectiveness was determined from the work histories. The fatigue model used this data to predict increases in human factors accident risk relative to chance and to determine if the relationship between effectiveness and accident risk was stronger for human factors than nonhuman factors accidents. This determined validity. Effectiveness scores were calibrated by associating the reduction in effectiveness with statistically reliable cumulative increases in accident risk.

Accident risk was defined entirely in terms of the Risk Ratio specified below:

$$\text{Risk ratio} = \frac{\text{Accidents at effectiveness level } x \text{ / Total number of accidents}}{\text{Work time at effectiveness level } x \text{ / Total work time}} \tag{7.1}$$

If the proportion of accidents occurring at any given level of effectiveness is approximately equal to the proportion of work time spent at that effectiveness level, then operator effectiveness is unrelated to accident occurrence (i.e., the accident risk should be close to one). If lower effectiveness is related to a greater likelihood of an accident, then accident risk should increase with decreasing effectiveness (i.e., be greater than one at lower effectiveness levels and less than one at higher effectiveness levels).

Put simply, FAST takes into account two major sources of information to predict cognitive effectiveness. This is illustrated in Figure 7.4.

Hursh et al. (2006, 2008) reported that human factors accidents displayed a reliable circadian rhythm, but nonhuman factors accidents did not. Their data is presented in Figure 7.5 which shows the risk of an accident on the left axis and the predicted inverse of cognitive effectiveness on the right axis as a function of time of day on the x-axis. The data have been double plotted to emphasize the circadian rhythm. For human factors accidents the correlation of time of day accident risk with the circadian pattern derived from the SAFTE model ($r = 0.71$, $p < 0.05$) is reliable. Circadian rhythms account for 51% of the variance in the time of day at which human factors accidents occur, while circadian rhythms only account for 6% of the variance ($r = 0.25$) in the time of day at which nonhuman factor accidents occur.

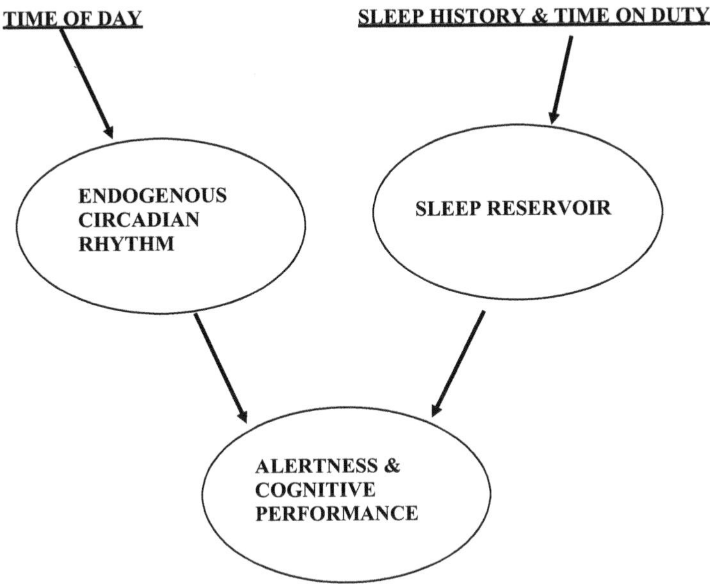

FIGURE 7.4   Simplified model of FAST. (After Raslear, 2014a.)

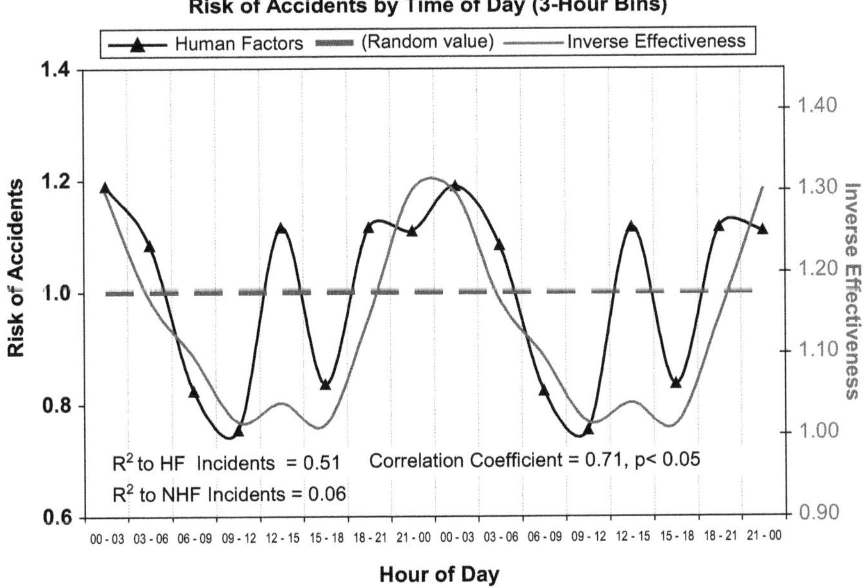

FIGURE 7.5   Risk of human factor accidents, predicted inverse effectiveness as a function of time of day. (After Hursh et al., 2006, 2008.)

The FAST software combines the effects of circadian rhythms with opportunities to sleep. Sleep fills the "sleep reservoir" and does so faster during normal sleep times. The reservoir is drained during awake periods (see Hursh et al., 2004 for details). The result is cognitive effectiveness. Hursh et al. (2006, 2008) hypothesized that human factors accident risk would increase as cognitive effectiveness decreased, and that there would be no relationship between effectiveness and nonhuman factors accident risk. Figure 7.6 shows human factors accident risk as a function of cognitive effectiveness.

Recall from our discussion of the Risk Ratio (RR), that if RR = 1, then effectiveness is not related to accident occurrence. Values of RR >1 signify an increase in accident risk, and values of RR <1 indicate a decrease in accident risk. For human factors accidents, RR has a value <1 when effectiveness is 90–100 (well-rested) and has a value of 1.65 when effectiveness is <50 (severely fatigued, see Raslear et al., 2011, Table 1). The correlation between RR and effectiveness is −0.93 ($p < 0.01$), which confirms the Hursh et al. (2006, 2008) hypothesis concerning human factors accidents. Hence, cognitive effectiveness accounts for 86% of the variance in accident risk. Note the dashed line in Figure 7.6 which represents the mean RR for nonhuman factors accidents. The correlation between RR and effectiveness for nonhuman factors accidents was not reliable ($r = -0.14, p > 0.05$). This confirms the nonhuman accident risk hypothesis. Together, confirmation of these two hypotheses constitutes validation of the SAFTE model and FAST software (see Raslear, 2013b for detailed discussion of criteria and procedures for validating and calibrating biomathematical models of fatigue).

Hursh et al. also sought to calibrate the SAFTE model. Calibration literally means to set the gradations on a measuring instrument such as a thermometer.

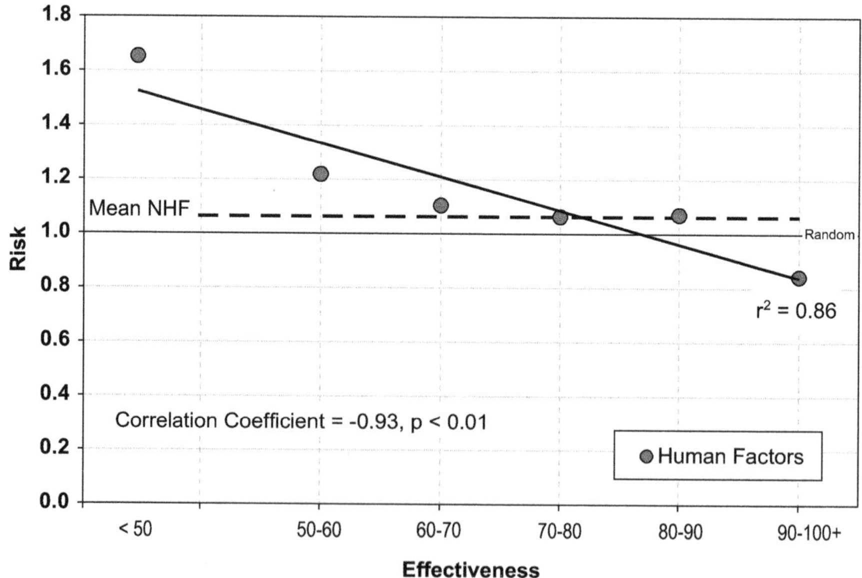

FIGURE 7.6 Human factors accident risk as a function of cognitive effectiveness. (After Hursh et al., 2006, 2008.)

In this instance, the most important gradation is the level of effectiveness below which a reliable increase occurs in accident risk (cumulative risk).[15] This is the threshold of unacceptable fatigue. Equally important are the effectiveness levels at which a person is considered non-fatigued and at which a person is considered severely fatigued.

To do determine these values, criterion levels were set to include the proportion of all accidents and the proportion of exposure time that occurred with an effectiveness score at or below that level. The ratio of these two proportions at each level was used to determine the effectiveness level below which an increased cumulative risk of accidents would exist. A ratio reliably greater than 1.0 at a particular criterion level indicates an increase in accident risk.

Table 7.3 summarizes the results and shows the percent of work time and accidents at each criterion level.

Table 7.3 shows that human factors accidents have a cumulative risk reliably greater than expected by chance for all effectiveness criterion scores below 90. However, effectiveness >90 is protective since there is a reliable 16% reduction in HF accidents. This defines the level at which a person is non-fatigued since the SAFTE model would never produce a score of <90 for a person who consistently gets 8 h of sleep each night.

Below a score of 90, there are small but reliable increases in accident risk. Fatigue risk gradually increases as effectiveness levels decrease. However, human factors accident risk is reliably greater than nonhuman factors accident risk and reliably greater than chance at effectiveness scores ≤ 70. This suggests that an effectiveness score of 70 might be a reasonable "threshold" for unacceptable fatigue. To further explore the reasonableness of 70 as a threshold, Hursh et al. examined human factors accident codes to determine if accidents with an effectiveness ≤ 70 were caused by errors that are typical of a fatigued state. If those accidents were caused by errors that are typical of a fatigued state, that would suggest that the relationship between effectiveness and human factors accident risk is meaningful and not due to extraneous factors.

TABLE 7.3

Human Factors Cumulative Risk by Criterion Effectiveness Levels

Criterion Effectiveness[a] Score	Human Factors Accident Risk (%)	Percent of Work Time	Human Factors Cases Number (%)[b]
>90	−16[c]	42	259 (35)
≤90	+11[c]	58	472 (65)
≤80	+14[c]	35	289 (40)
≤70	+21[c,d]	19	166 (23)
≤60	+39[c,d]	7	71 (10)
≤50	+65[c,d]	2.7	33 (4.5)

Source: After (Hursh et al., 2006, 2008).

[a] Effectiveness at accident time based on 30-day work histories processed using the SAFTE fatigue model.

[b] HF cases (two crew members per accident) in 2½ years, excluding consistent night workers. The percentages >90 and ≤90 sum to 100%. The percentages ≤90 are cumulative and do not sum to 100%.

[c] Significantly different from chance ($p < 0.05$).

[d] Significantly different from nonhuman factors accident risk.

Table 7.4 shows the 10 most over-represented human factors cause codes with effectiveness ≤70. In most of the approximately 400 human factors accidents two crew members were involved. That resulted in 880 cases because an accident could have more than one cause code assigned to it. Of the 880 cases, 215 had an effectiveness score ≤70. The proportion of all human factors cause codes with an effectiveness score ≤70 was 0.244 (215/880). Table 7.4 shows the total frequency and the frequency with an effectiveness score ≤70 for each individual cause code. Relative likelihood is the proportion of each cause code with an effectiveness score ≤70 relative to the total frequency of that cause code divided by the proportion of all human factors cause codes with an effectiveness score ≤70. For example, in Table 7.4 cause code 701 has a proportion of 1.0 of accidents with an effectiveness score ≤70 (4/4). The relative likelihood of this cause code having an effectiveness score ≤70 is 4.09 (1.0/0.244).

For all of the cause codes listed in Table 7.4, the proportion of these accidents with an effectiveness score ≤70 was greater than the overall proportion of all human factors accidents with an effectiveness score ≤70.

Most of the cause codes in Table 7.4 are associated with human errors that would be expected to increase with fatigue (Raslear, 2013). Many involve possible lapses of attention (H222, H221, H605; Tilley and Brown, 1992) or failures to complete all actions required in a task (H701, H019, H311; Craig & Cooper, 1992). Hursh et al. argued that

---

TABLE 7.4

Human Factors Cause Codes with an Effectiveness ≤70 That Are Over-Represented

Rank	Cause Code	Frequency with Effectiveness ≤70	Total Frequency	Relative Likelihood	Cause Code Description
1	H701	4	4	4.09	Spring switch not cleared before reversing
2	H311	5	7	2.92	Moving cars while loading ramp, hose, or chute not in proper place
3	H019	8	15	2.18	Failure to release hand brake on cars
4	H222	4	8	2.05	Automatic block or interlocking signal displaying other than stop indication – failure to comply
5	H404	7	17	1.69	Train order, etc. failure to comply
6	H504	4	10	1.64	Buffing or slack action excessive, train makeup
7	H221	12	31	1.58	Automatic block or interlocking signal displaying stop indication – failure to comply
8	H605	18	53	1.39	Failure to comply with restricted speed in connection with the restrictive indication of a block or interlocking signal
9	H399	4	12	1.36	Other general switching rules
10	H503	20	60	1.36	Buffing or slack action excessive, train handling

Source: After Hursh et al. (2006, 2008).

this provided meaningfulness to the adoption of the ≤70 threshold. Meaningfulness is defined in a number of ways in the measurement literature (see for example, Falmagne, 1985; Luce et al., 1990). According to Falmagne (1985, p. 343), "…a model is a convincing explanation only if it is physically realizable." Connecting a physical error that contributed to an accident to a fatigue score provides meaningfulness.

In epidemiology, risk is determined by both the magnitude and the duration of exposure to an agent (Lilienfeld and Lilienfeld, 1980). Thus far we have considered fatigue severity but not duration of exposure to fatigue. Looking again at Table 7.3, we see that at effectiveness ≤70 the duration of exposure (work time) is 19%. "This suggests that greater than 20% of work time at an effectiveness level below 70 is a good first-order approximation of severity and duration exposure for establishing a fatigue benchmark for rail operations. A reasonable goal for fatigue risk management would be to keep employee work time within that threshold." (Raslear et al., 2011, p. 164). This is the threshold stipulated for FAST in the HOS regulations published on October 15, 2011 (49 CFR 228; Federal Register Number 2011–20290) for passenger service employees (see also Raslear, 2013b).

Table 7.5 shows the more detailed calibration of effectiveness scores relative to hours awake and suggested nomenclature for gradations of fatigue (Raslear et al., 2011; Raslear, 2013). It is of note that, in the laboratory (Hursh et al., 2004, Figure 7), that an effectiveness of 94 is associated with a lapse likelihood of 1.0 (indicating that the risk of a lapse is neutral), while at an effectiveness of 70 the likelihood of a lapse is 5 (indicating a 5-fold increase in risk). Performance on a driving simulator at an effectiveness of 70 is equivalent to a blood alcohol concentration 0.08 (Arnedt et al., 2001).

Biomathematical fatigue models predict opportunities for sleep associated with a work schedule. The accuracy of the predictions depends, in part, upon the accuracy of the sleep estimation. The SAFTE biomathematical model of fatigue estimates the amount of sleep that is obtained by using an algorithm called AutoSleep. Research sponsored by FRA not only examined the validity and calibration of SAFTE, but also the accuracy of the sleep estimates produced by FAST which contains AutoSleep (Federal Railroad Administration, 2011a; Gertler et al., 2012). These two studies

TABLE 7.5

Effectiveness Scores as a Function of Hours Awake and Nomenclature

Effectiveness Score	Hours Awake (HH:MM)	Nomenclature
98	14:00	Not fatigued
94	15:10	Not fatigued
90	16:00	Not fatigued
80	18:00	Mildly fatigued
77	18:30	Moderately fatigued
70	21:00	Moderately fatigued
69	22:00	Very fatigued
60	40:50	Very fatigued
50	42:30	Severely fatigued

compared AutoSleep estimates based on work/rest diaries with actigraphy[16] or sleep reported in work/rest diaries.

FRA (2011a) used data previously collected in the diary studies from T&E employees, dispatchers, signalmen, and MOW employees (Gertler and Viale, 2006a,b, 2007; Gertler and DiFiore, 2009). Employees with irregular day and night work were analyzed separately from those with at least 50% of work starts between 2200 and 0400 (regular night workers). The results are shown in Table 7.6. The accuracy of AutoSleep predictions varies from a low of 79% to a high of 92%. Overall, with the exception of night dispatchers, there is a bias for AutoSleep to underestimate sleep durations. This would bias fatigue estimates to higher levels of fatigue. For regular night workers, fatigue estimates were consistently higher (i.e., effectiveness scores were lower) for AutoSleep than for diary sleep. Effectiveness scores were approximately 5 points lower for AutoSleep. For irregular day and night workers, the estimates were almost equal (see FRA, 2011a, Figure 4).

The Gertler et al. (2012) study collected actigraph records of sleep and wake periods along with daily diary data from 41 locomotive engineers working in freight service over a 14 d period. They computed the percentage of time intervals when the actigraph and diary showed the participant in the same sleep state (agreement). Then, signal detection theory (SDT, see Chapter 5), was used to determine the extent and direction of the bias, or systematic errors, in AutoSleep and the sensitivity of AutoSleep to detect sleep periods. Depending on the settings used for AutoSleep (see Gertler et al., Table 5.), the agreement ranged from 84.65% to 86.94% with an error that ranged from −13 min to +6 min The actigraph hours of sleep and the AutoSleep predictions were not reliably different ($p > 0.05$). The SDT analysis found that the sensitivity (d') of AutoSleep ranged from 1.85 to 2.09 (equivalent to 82% and 85% agreement in an unbiased procedure). Bias (β) ranged from 1.74 to 1.89, indicating a bias for AutoSleep to say "awake". β is determined, in part, by the prior odds of the event being detected. In this case, the event is "awake". Gertler et al. found that the prior odds (p(awake)/p(asleep)) ranged from 2.40 to 2.56, predicating a bias to say "awake". This is not surprising considering that the participants in the study slept less than 7 h according to actigraph data which would mean that p(asleep) = 0.29 and

---

TABLE 7.6

Accuracy of Sleep Predictions

	Irregular Day and Night Worker				
	Signalman	**MOW**	**Dispatcher**	**T&E**	**Mean**
Mean Agreement	92%	92%	90%	88%	90.5%
Error (estimate − diary)	−24 min	−21 min	−3 min	−10.8 min	−14.7 min
	Regular Night Worker				
	Night Dispatcher		**Night T&E**		**Mean**
Mean Agreement	79%		82%		80.5%
Error (estimate − diary)	+18.8 min		−40.2 min		−10.7 min

Source: After FRA (2011a).

p(awake) $= 1 - p$(asleep) $= 0.71$ so that the prior odds (based on actigraphy) $= 2.45$. This means that AutoSleep parameters can be adjusted to make the algorithm less biased (see Gertler et al. for more detail).

Although Raslear (2013b) lists six biomathematical models of fatigue that meet the basic requirements to satisfy criterion 1 for validation, only one other model has been validated by the FRA for use in the railroad industry to analyze work schedules. Elements of railroad management requested that FRA formally validate the Fatigue Audit InterDyne model [FAID (Roach et al., 2004)]. The concern was that if SAFTE was the only validated model, a monopoly would be established to the detriment of the financial interests of the railroad industry. Consequently, FRA validated FAID using the database that was used to validate SAFTE (Tabak and Raslear, 2010). The Tabak & Raslear report provides regression equations to allow a translation between FAST effectiveness scores and the metrics of FAID.

The Fatigue Status of the U. S. Railroad Industry

Given the ability to measure fatigue, the Human Factors Research program at FRA set out to

Understand the size and scope of the fatigue issue in railroads
Understand how to manage and mitigate fatigue
Reduce fatigue-related accidents.

The ultimate goal of the fatigue research program was to provide a scientific basis for HOS regulations.

A series of diary studies was performed to examine work schedules and sleep patterns for the major safety-critical jobs in the railroad industry. These included:

Train & Engine Employees (Gertler and DiFiore, 2009);
Passenger Train & Engine Employees (Gertler and DiFiore, 2011);
Signalmen (Gertler and Viale, 2006a);
Dispatchers (Gertler and Viale, 2007);
Maintenance of Way Employees (Gertler and Viale, 2006b).

With the exception of the Maintenance of Way Employees (MOW), all of these labor groups are covered under HOS regulations. Consequently, MOW workers are a useful comparison group for the other diary studies. In the data collected from these groups there were no reported accidents. The database for the SAFTE validation does contain information related to accidents and can form a comparison group for the diary study, Train & Engine employees.

As previously noted, we know that fatigue is largely a function of sleep and circadian rhythms. The diary studies show that sleep is largely a function of work schedules. The basic work schedule characteristics that affect sleep are work duration, time of day of work, and schedule variability (predictability). Consequently, fatigue exposure is largely determined by work schedules. Furthermore, fatigue risk and the probability of human factor accidents are determined by fatigue exposure. The following material provides a brief summary of the diary studies. Two detailed summaries of the diary and other pertinent studies can be found in Gertler et al. (2013) and Raslear et al. (2013).

Table 7.7 shows the 12 types of work schedules that the diary studies found in the railroad industry. Dispatchers and T&E employees work round-the-clock. Signalmen and MOW workers work primarily during daytime but are often called after regular shifts to perform work on an emergency or accident. Extra board dispatchers, most road freight crews, and extra board T&E in passenger service do not have fixed start times while the other jobs do.

Since diaries were kept for a 2-week period, (Gertler et al., 2013; Raslear et al., 2013), many summary statistics are reported for 2-week periods. The average (median) work duration (h:min) for passenger T&E was 82:31, for signalmen 82:43, for dispatchers 82:59, for MOW 84:04, and for T&E 85:15. These averages do not show much variation across the 2-week period

Railroad work occurs around-the-clock, 7-days-a-week. However, as Figure 7.7 shows, operations are not uniform throughout the day. Signalmen and MOW have very similar work schedules. Dispatchers have a clear three shift work pattern. The T&E groups appear to have a nearly uniform pattern of work, with the Passenger T&E group working more during the day time. However, none of the five groups follow a uniform distribution of start and end times; all had χ^2, $p < 0.05$.

Schedule variability can disrupt a worker's normal circadian sleep pattern and lead to fatigue. Prior studies (AAR, 1994; GAO, 1992; Pollard, 1991, 1992) examined schedule variability in terms of start time variability (STV). A change in start time of more than 1 h from the previous day was the definition for STV in the diary studies for signalmen, MOW and dispatchers.

Most dispatchers ($\approx$ 70%) work the same shift and have no STV (mean STV = 0.25 h). However, 19% of relief dispatchers (mean STV = 3.03 h) and of 12% extra board dispatchers (mean STV = 1.68 h) worked various shifts in a forward

---

TABLE 7.7

Types of Work Schedules in the Railroad Industry

Schedule	Signalmen	Maintenance of Way	Dispatcher	Train & Engine Personnel	Passenger T&E
4-day	√	√			
5-day	√	√			
8/6	√	√			
1st Shift			√		
2nd Shift			√		
3rd Shift			√		
Relief			√		
Extra board			√	√	√
Variable start time				√	
Fixed start time				√	
Straight Thru					√
Split					√

Source: From Raslear (2014a).

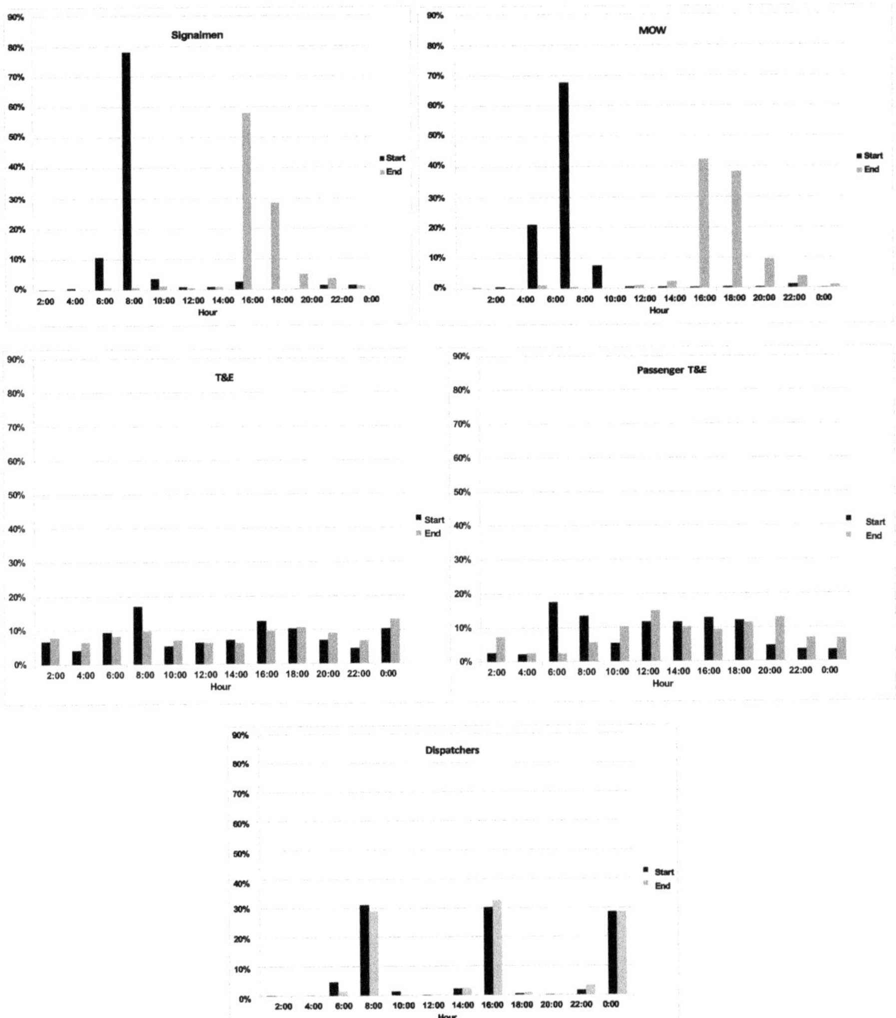

FIGURE 7.7　Work start and end times for railroad jobs. (After Raslear et al., 2013.)

rotating pattern (HOS regulations do not allow backward rotating schedules for dispatchers; see also Pilcher & Coplen, 2000).

Among signalmen involved in construction jobs, 10% experienced STV (average 0.23 h), while 37% of those involved in non-construction jobs experienced STV (average 0.5 h). Similar data were found among MOW workers. 16% of MOW workers in production jobs had STV = 1.02 h, and 22% in non-production jobs had STV = 1.5 h.

In the T&E diary studies STV was examined in terms of the difference in start time from the prior work period (vs. the previous day). As should be apparent from Figure 7.7, T&E jobs had the greatest variability: fixed start time jobs had 3.3 h mean STV, and variable start time jobs had 7.1 h mean STV. Overall, 25% of T&E

jobs had STV of 9 h or more. By contrast, passenger T&E jobs had much less STV (mean = 1.1 h), probably because they work on scheduled trains.

In meetings about HOS railroad labor often talked about the role of the unpredictability of work schedules in the industry in causing employee fatigue. They would say that unpredictable work schedules prevent employees from planning personal activities and sleep, resulting in fatigue. Obviously, this is related to STV, and Raslear (2014b) took a statistical approach to the problem of determining the predictability of work start times for T&E freight and passenger employees.

Raslear (2014b) generated start time difference distributions by calculating the difference in start times between successive shifts, regardless of intervening non-work days. For example, if an employee starts work on Tuesday at 0800 and at 1000 on Thursday (the next work start), the start time difference is +2 h. If an employee starts on Tuesday at 0800 and the next work start is at 0600, the start time difference is −2 h. Multiple starts in the same day are treated the same way.

Three previously published FRA databases were used:

The Fatigue Accident Validation database (FAV, http://www.fra.dot.gov/eLib/details/L03039, FRA, 2011b); the Work Schedules and Sleep Patterns of Train and Engine Service Workers database (http://www.fra.dot.gov/Page/P0604, FRA 2008); the Work Schedules and Sleep Patterns of Passenger Train and Engine Service Workers database (http://www.fra.dot.gov/Page/P0604, FRA 2008).

The FAV provided data for freight T&E employees on days with human factors (HF) accidents and on days preceding HF accidents. The other two databases provided data for freight and passenger T&E employees who were not involved in an accident during the data collection period. Table 7.8 shows the sample sizes from which start time difference distributions were generated.

Raslear (2014b) hypothesized that the variance (σ^2) of the distributions would be ordered as

$$\sigma^2_{\text{Accidents}} > \sigma^2_{\text{Pre-accident}} > \sigma^2_{\text{Freight no accidents}} > \sigma^2_{\text{Passenger no accidents}} \tag{7.2}$$

The variance provides a measure of the unpredictability of start times from day to day relative to the expected value of the distribution or mean (Hays, 1963). The variance would be zero only if every start time difference was exactly the same as the mean. Hence, a smaller σ^2 indicates greater predictability.

TABLE 7.8

Start Time Difference Distributions and Sample Sizes

Distribution	Sample Size
Freight T&E, day of accident	704
Freight T&E, days preceding accidents	11,876
T&E (93% freight), days without accidents	2,412
Passenger T&E, days without accidents	3,347

Note: A small number of passenger service employees (7% of cases) are included in this database.

Figure 7.8 shows the start time variances for the four distributions. The order hypothesized by Raslear (2014b) is apparent. Statistically, $\sigma^2_{\text{Accidents}} > \sigma^2_{\text{Pre-accident}}$ (F703, 11,875 = 1.09, $p < 0.05$); $\sigma^2_{\text{Pre-accident}} > \sigma^2_{\text{Freight no accidents}}$ (F11876, 2,411 = 1.08, $p < 0.05$); $\sigma^2_{\text{Freight no accidents}} > \sigma^2_{\text{Passenger no accidents}}$ (F2411, 3,347 = 1.74, $p < 0.05$). Moreover, there is a reliable correlation ($r = 0.9857$, $p < 0.05$) between fatigue (percent of work time at a FAST score ≤ 90) and the standard deviation (σ) of start time difference for the four distributions.

Work schedules and their associated variability provide sleep opportunities but are not the exclusive determinants of the amount of sleep that workers obtain. Gertler et al. (2013) examined daily primary sleep, total daily sleep, and the average number of sleep periods with one-way ANOVAs. They defined daily primary sleep as the longest sleep period ending on a given calendar day. Total daily sleep is the sum of all sleep periods that end on a calendar day. Each sleep period ending on a calendar day is counted to determine the number of sleep periods per day. Average daily sleep periods determine the extent to which an employees' sleep is segmented. Segmentation can result in fatigue if sleep periods are not sufficient to allow individuals to adequately cycle through the typical stages of sleep.

The 12 schedule types were significantly different from one another with regard to daily primary sleep $F(11, 1{,}573) = 31.39$, $p < 0.0001$), total daily sleep ($F(11, 1573) = 22.35$, $p < 0.0001$), and the average number of daily sleep periods ($F(11, 1{,}573) = 14.75$, $p < 0.0001$). Figure 7.9 shows the total daily sleep by type of schedule. Referring back to Table 7.7, we can see that fixed start time, variable start time, straight and split schedules are T&E freight and passenger schedules. Extra board also includes dispatchers who have relief, 1st, 2nd, and 3rd shifts schedules. 4-day, 5-day and 8 days on/6 days off schedules are worked by MOW and signalmen. Third shift and relief schedules obtain the least amount of total daily sleep, while split and straight assignment employees obtain the most amount of total daily sleep. Employees working third shift, relief, and split assignments report the highest number of daily sleep periods, but third shift and relief employees do not approach the total amounts of daily sleep that split assignment workers are able to obtain.

In 2004 FRA issued Safety Advisory 2004-04 concerning sleep disorders and safety critical employees (Federal Railroad Administration, 2004). The sleep diary studies also obtained information about sleep disorders in the five employee groups. The type of sleep disorder, however, was not specified for the MOW and signalmen

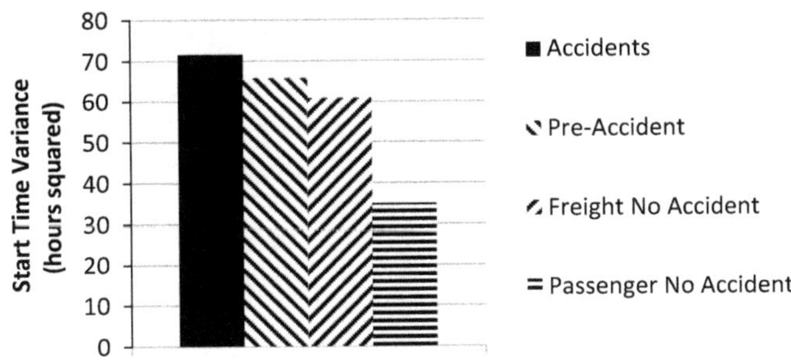

FIGURE 7.8 Start time variances. (After Raslear, 2014b.)

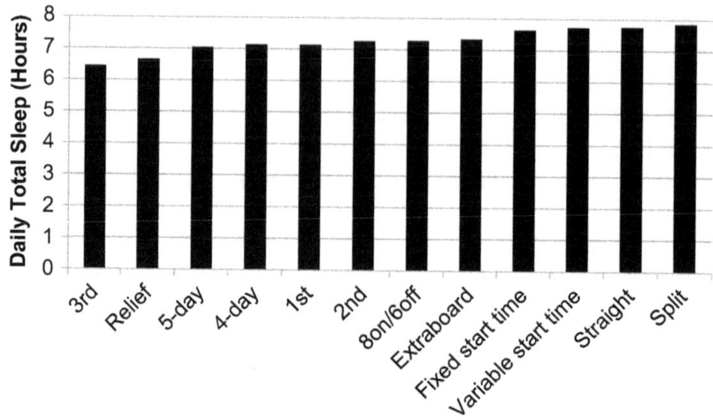

FIGURE 7.9   Total daily sleep by type of schedule. (After Gertler et al., 2013.)

groups which reported a prevalence of unspecified sleep disorders as 5.7% for signalmen, and 6.7% for MOW. Table 7.9 (Raslear, 2014c) shows the prevalence of sleep apnea for dispatchers, freight T&E employees, passenger T&E employees and the general U. S. male population.

From Table 7.9 it is clear that the prevalence of sleep apnea in dispatchers and T&E employees is nearly two times higher than the general U. S. male population. This comparison is apt, since only 9.4% of the sampled groups were female. Between 20% and 34% of the railroad employees do not receive treatment for sleep apnea. It is possible that the figures in Table 7.9 underestimate the prevalence of sleep apnea in these groups because many may not be aware of their condition. Raslear (2014c) notes that sleep apnea has been implicated in several high visibility railroad accidents and has been the subject of five NTSB safety recommendations. In view of this, it is interesting that FRA has withdrawn a proposed regulation (Federal Motor Carrier Safety Administration and Federal Railroad Administration, 2016) to require T&E employees to undergo testing for sleep apnea (Federal Motor Carrier

TABLE 7.9

Sleep Apnea Prevalence in Safety-Critical Railroad Jobs

Group	Number of Participants	Number with Sleep Apnea	Percent with Sleep Apnea [Mean (95% CI)]	Percent Receiving Treatment
Freight T&E	250	21	8.40 (8.34–8.46)	66
Dispatchers	443	33	7.45 (7.42–7.48)	66
Passenger T&E	256	15	5.86 (5.82–5.90)	80
Total	949	69	7.27 (7.26–7.28)	69.6
U.S. males			4.0	

Source: After Raslear (2014c).

Safety Administration and Federal Railroad Administration, 2017). Subsequent to the withdrawal of the proposed regulation the "National Transportation Safety Board ... determined that two commuter railroad terminal accidents in the New York area were caused by engineer fatigue resulting from undiagnosed severe obstructive sleep apnea." (National Transportation Safety Board, 2018). Clearly the *status quo* is not addressing this issue.

Obviously, the interplay between work schedules, schedule variability, circadian rhythms, and sleep patterns is complex and difficult to summarize in a straight-forward fashion. That is the function of biomathematical models of fatigue. Figure 7.10 shows the fatigue exposure of six distinct groups of railroad employees: those who had accidents, freight T&E workers, dispatchers, MOW workers, signalmen, and passenger T&E workers. Fatigue exposure is the proportion of work time spent in a level of fatigue as indicated by effectiveness (FAST) scores.

Employees who have accidents have the greatest exposure to the highest level of fatigue (Very Fatigued and Fatigued), followed by freight T&E and dispatchers. By comparison, passenger T&E, MOW and signalmen have the least fatigue exposure and the highest proportion of work time Not Fatigued.

Figure 7.11 shows the mean FAST scores for accidents and work groups. Here the same pattern emerges. The least fatigued group is passenger T&E, followed by MOW and signalmen. Freight T&E and dispatchers are nearly equivalent, and the most fatigued group is the employees who had accidents.

What is the relative contribution of sleep and circadian rhythms to fatigue? Going back to Figure 7.4, we proposed that fatigue was a function of sleep and circadian rhythms. Expressed mathematically, Fatigue = f (sleep + circadian rhythm). Hursh et al. (2006, 2008) found that FAST scores (fatigue) accounted for 86% of the variance in relative risk of an accident (see Figure 7.6). Hursh et al. also found that circadian rhythms accounted for 51% of the variance in relative risk of an accident

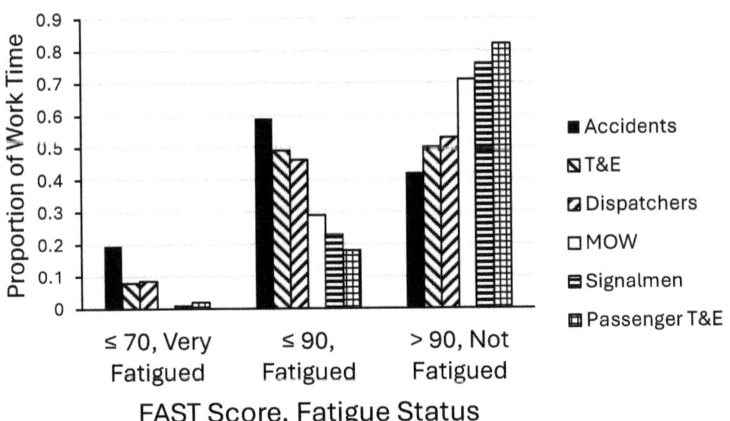

FIGURE 7.10 Fatigue status for accidents and work groups. (After Gertler et al., 2013.) From left to right the groups are accidents, T&E, dispatchers, MOW, signalmen, passenger T&E.

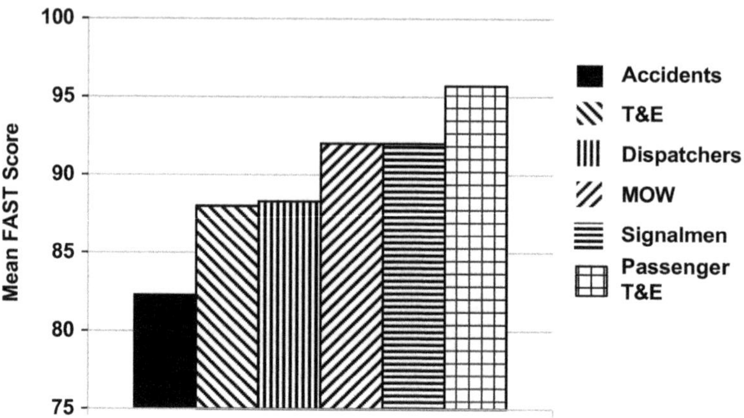

FIGURE 7.11   Mean FAST scores for accidents and work groups. (After Gertler et al., 2013.)

(see Figure 7.5). That implies that sleep should account for approximately 35% of the variance (86 − 51 = 35). Gertler et al. (2013) determined that total sleep duration accounted for 34% of the variance in FAST scores (see Figure 7.9). Gertler et al. (pp. 58–59) conclude that this "…is consistent with the expectation, based on the scientific literature, that fatigue is a function of sleep and circadian influences. Sleep duration is obviously an important aspect of fatigue exposure and risk, but it is not the only factor. Circadian rhythms must also be considered, and this is the true value of biomathematical models in fatigue management: they allow for easy evaluation of the interaction of multiple processes."

Hursh et al. (2006, 2008) examined the relative contributions of sleep and circadian rhythms in T&E employees which is the basis for the conclusion that circadian rhythms account for 51% of the variance in the relative risk of a human factors accident. Calabrese, Mejia, McInnis, France, Nadler, and Raslear (2017) examined time of day effects in fatal and nonfatal injuries in MOW workers and signalmen. This study found that circadian models of odds ratios of nonfatal injuries accounted for between 73% and 85% of the variance. A circadian model for fatal injuries accounted for 95% of the variance. These differences between the Hursh et al. and Calabrese et al. studies could be due to numerous factors. Human factors accidents can include nonfatal and fatal injuries, but injuries may be exclusive of accidents involving property damage. Consequently, circadian rhythms may account for more of the variance in injuries. Hursh et al. define relative risk (see Hursh et al., 2008, p. 10) differently than odds ratio is defined by Calabrese et al. Relative risk is the proportion of accidents at a time of day divided by the proportion of work time at a time of day. By contrast the odds of injuries at a time of day was obtained by dividing the number of accidents at that time by the number of accidents *at all other times of day*. The odds of an employee working hours at a time of day was obtained by dividing the number of worker hours in that bin by the total number of worker hours *at all other times of day*. The odds ratio is the ratio of these two odds (see Ahlbom, 1993, p. 76). Odds are not proportions. Clearly, the relative contributions of sleep and circadian rhythms to fatigue and accidents is not settled science.

The Risk of a Human Factors Accident Due to Fatigue Exposure

Gertler et al. (2013) define the risk of a human factors accident given fatigue (HFA|E, where E is Effectiveness) as the probability of a human factors accident given fatigue multiplied by the cost of that accident:

$$\text{Risk} = p(\text{HFA}|E) \times \text{Cost}$$

This is the engineering definition of risk (for other definitions see Gertler et al., p. 60–62) and is most appropriate for use in discussing risk in the railroad industry. Risk defined this way is also the expected value of a decision.

Gertler et al. (p. 63) describe the process of determining the value of $p(\text{HFA}|E)$ for values of E. The end result is shown in Table 7.10 which shows $p(\text{HFA}|E)$ per 200,000 employee hours.[17] When Effectiveness is ≤ 50 (severely fatigued), the probability of an HFA is almost double that when there is no fatigue ($E > 90$). Table 7.9 also shows the average cost of HFA as a function of E (Hursh et al., 2011). Accidents that involve fatigue cost much more than accident in which there is no fatigue ($E > 90$). The average cost of an HFA when E is between 71 and 90 is nearly double the cost when there is no fatigue. When $E \leq 70$, the average cost is four times the cost with no fatigue. Risk ($p(\text{HFA}|E) \times \text{cost}$) as a function of E accelerates even more quickly. For $71 \leq E \leq 90$, risk is nearly 2.5 times the cost with no fatigue. When $E \leq 70$, risk is 5.6 times the risk with no fatigue.

What Is the Impact of RSIA?

RSIA was passed in 2008. Raslear (2013a) collected data on accidents and fatigue from before the passage of RSIA (2003–2005) and after the passage of RSIA (2010–2012). These are both 3 year time periods which makes the comparison of accident frequencies commensurate. Figure 7.12 shows the mean FAST scores for freight and passenger railroads for the Hursh et al. (2008) diary study and for accidents in 2003–2005 and 2010–2012. The FAST diary study data are for non-accidents and provide a comparison with the accident data. As expected, the non-accident data indicate less fatigue than the accident data. The post-RSIA data also show a trend towards less fatigue than the pre-RSIA data, but the trend is not statistically reliable.

Figure 7.13 examines the accident data in more detail by plotting the proportion of accidents in each time period as a function of FAST score. For FAST Scores from ≤ 50 (Severely Fatigued) to 61–70 (Very Fatigued), the proportion of accidents is greater for the pre-RSIA period (2003–2005). The proportion Not Fatigued (>90) is

---

TABLE 7.10

$p(\text{HFA}|E)$, Average Cost and Risk as a Function of Effectiveness (E)

Effectiveness (E)	≤ 50	≤ 70	71–90	>90
$p(\text{HFA}|E)$	0.276	0.210	0.188	0.152
Average cost		$1,589,938	$771,686	$395,283
Risk		$334,518.16	$145,163.24	$59,968.60

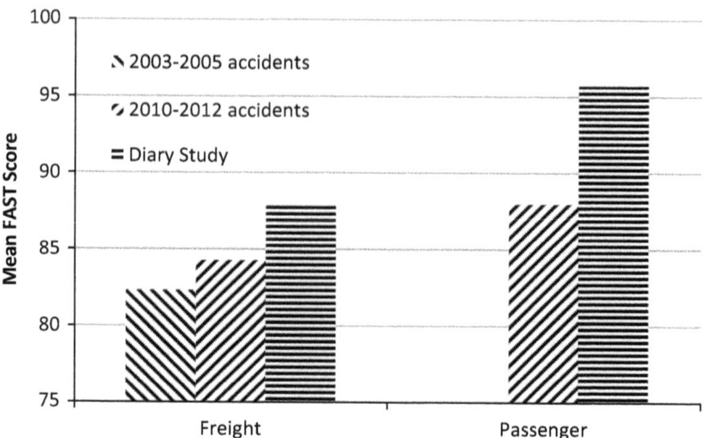

FIGURE 7.12 Mean FAST scores for freight and passenger accidents before (2003–2005) and after (2010–2012) RSIA. Non-accident data from Hursh et al. (2008) is presented for comparison. (After Raslear, 2013a.)

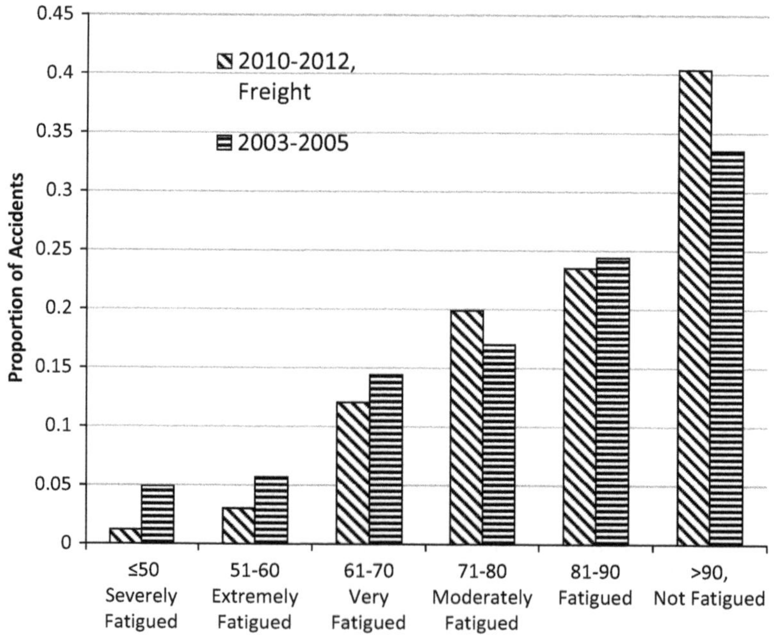

FIGURE 7.13 Proportion of freight accidents before and after RSIA as a function of FAST score. (After Raslear, 2013a.)

greater for the post-RSIA period (2010–2012). The trend is again in the right direction, but the differences are not reliable. Two to four years post-RSIA may not have provided sufficient time to assess the effect of RSIA.

The Future

The Human Factors Research Division of FRA has given FRA and the railroad industry the tools to measure and manage fatigue in safety critical employees. Raslear (2014a) in an address to the FRA, railroad labor and management outlined what these tools are. There are two databases (Work Schedules and Sleep Pattern Survey Data, Federal Railroad Administration, 2008; The Fatigue Accident Validation Database, Federal Railroad Administration, 2011b) that provide a baseline from which to ascertain whether there have been changes in the fatigue status of the U. S. railroad industry. There is a validated and calibrated tool for measuring fatigue (FAST). The FAST tool should be used to continuously assess fatigue in HF accidents and to develop a database of accidents that can be compared to the Fatigue Accident Validation Database. In addition, periodic work/rest surveys should be conducted to assess non-accident fatigue, using FAST, in the industry by comparison to the Work Schedules and Sleep Pattern Survey Data. This is the primary way by which changes in regulations, policies and practices can be evaluated regarding the future status of fatigue in the railroad industry. You can't manage what you don't measure.

Railway Worker Fatigue and Technology

Fatigue Website

The Railroaders' Guide to Healthy Sleep (https://railroadersleep.fra.dot.gov/) was a project conceived by Mike Coplen to provide scientifically valid information on sleep to the railroad employees, and to provide free preliminary screening for sleep apnea. This website is owned and funded by the Federal Railroad Administration. It was originally created by the Division of Sleep Medicine at Harvard Medical School in 2012 and updated by the U.S. DOT Volpe Center in 2015.

The FRA Human Factors Research Division considered railroaders an at-risk population for fatigue and sleep disorders because of the demographics of this group. A needs assessment was conducted to better understand the railroader target audience. They are

98% male;
59% ≥45 years of age;
53% conductors; 39% locomotive engineers;
82% work schedules that "mostly change" week to week;
36% high school/GED or vocational/trade school; 38% some college;
57% spend <10 h per week online;
36% had not used web to find medical information in last 6 months.

FRA partnered with the Volpe Center to construct a website of high professional quality. Internationally recognized experts from the Harvard Medical School Division of Sleep Medicine provided impartial scientific and health information for the website. The WGBH Educational Foundation was engaged to provide expertise on the layout and presentation of the materials on the website. Railroad employees were interviewed, consulted, and filmed in real-life situations that were featured on the website to provide realism. Usability testing was conducted throughout the development stage to ensure ease of access of the material. Terminology and jargon used in the industry was incorporated in reading materials that were appropriate for the education level of the audience. Stakeholder feedback was solicited from industry management and labor to ensure stakeholder support.

The goals of the website were to provide scientifically valid content on the importance of sleep. However, information about sleep would not alone solve the real-world challenges of balancing work and life in situations where work schedules are often unpredictable. So another goal was to provide proven, practical strategies to address work/life balance in the railroad industry. To do this the website needed to provide personal tools to address sleep-health issues such as a tool to anonymously assess the individual's risk for sleep disorders and a sleep-wake diary to track sleep patterns over time. The ultimate goal was to motivate railroaders to adjust sleep behavior in the aspects of their lives over which they had control.

The topics covered on the website include "Explore Sleep", which contains modules that discuss the basics of sleep science including how much sleep is needed and circadian rhythms. There is also a module called "Your Sleep Toolkit" which provides a sleep-wake diary, and tools to self-test sleepiness and reaction time. The topic "Improve Sleep" provides tips for managing sleep, naps, the use of caffeine and exercise. Also, in this module is information about sleep disorders and their treatment, and access to the anonymous sleep disorders screening tool. The topic "Balance Work and Life" provides modules with tips for the railroader to use to get better sleep and have a better quality of social life, for family members to support the railroaders' in meeting that goal, and advice from the railroad industry about the shared responsibility to manage fatigue. There is a topic "Screen Yourself for Sleep Disorders" which answers questions about the screening tool and provides access to the tool. The tool screens for obstructive sleep apnea, insomnia, restless leg syndrome, shift work sleep disorder, and narcolepsy. Individuals receive notification explaining their sleep disorder risk based on responses to the screening tool. There is a link to a zip-code lookup feature that allows users to locate qualified sleep specialist in their area. Finally, there is a topic "Shareable Resources" which provides short educational materials, promotional materials, and self-assessment tools.

Alertness Monitoring

Alertness is critically important in the operation of trains. A lapse in alertness of just a few seconds can result in a locomotive engineer not seeing a wayside signal indicating that the train must stop at a signal forward of the train. Trains often exceed a mass of 10,000 tons and are difficult to stop because the steel-wheel to steel-rail interface provides very low friction. Stopping distances of a mile or more are not uncommon. Although lapses are associated with fatigue and increase in frequency as fatigue increases (Hursh et al., 2004), lapses also occur with alcohol and drug (over-the-counter, prescription and illicit) use, illness, environmental conditions (heat, cold, noise, vibration), and the railroad operating environment: "Railroad work is characterized by large variations in cognitive workload. Both extended low-activity periods and moderate periods of high workload can occur in the same run. There is a strong element of repetitiveness that can breed complacency. Locomotive engineers often operate over the same territory over years or decades. Because the locomotive engineer is likely to encounter the same signal conditions run after run, a change in a critical signal may go undetected." (Sussman and Raslear, 2008, p. 165).

Efforts to ensure that locomotive engineers are alert and alive have a long history in railroading (Devoe and Abernathy, 1977). Deadman switches, a switch that the engineer must keep depressed with his or her foot, were an early form of device to ensure that the operator of the train was alive and functioning. Such devices are easily foiled (for instance, a weight can be placed on the switch), however, and devices that depend on active engineer behaviors were developed. 49 CFR § 229.140 – Alerters describes the functioning of alerters found in many locomotive cabs. Alerters have a timing cycle that depends on the speed of the train. The faster the train is moving the shorter the timing cycle (the timing cycle specified in seconds = 2,400 ÷ track speed specified in miles per hour). Alerters provide an audio alarm upon expiration of the timing cycle interval. This is preceded by a visual signal at least 5 s prior to the audio alarm. If a reset button is not pressed before the timing cycle expires, the locomotive brakes are applied to stop the train. Other active behaviors automatically reset the timing cycle. These include movement of the throttle handle, movement of the dynamic brake control handle, movement of the horn activation handle, movement of the bell activation switch, movement of the automatic brake valve handle, bailing the independent brake by depressing the independent brake valve handle. These are actions routinely performed by the locomotive engineer in controlling the train. However, anecdotally one hears that it is possible to devise mechanical methods to foil the alerter, and it is often said that locomotive engineers develop automaticity (pressing the reset button repetitively at an interval sufficient to avoid activation of the video or auditory alarm without conscious effort), thereby circumventing the purpose of the alerter. Consequently, FRA and other railroad safety entities have sought to improve the design of alerters or replace them with devices that directly detect alertness and prevent lapses (Centre for Sleep Research, 2003; Halcrow Pacific, 2006; Riley, 2005; Whitlock, 2002; Wright et al., 2007;). The devices range from those that measure physiological functions such as eye-based systems (e.g., FIT2000, PERCLOS), electroencephalogram (e.g., ABM Drowsiness Monitoring Device), or galvanic skin response (e.g., Engine Driver Vigilance Telemetric Control System) to those based on physical actions such as head droop (StayAlert). Cognitive alerters, devices like the alerters currently in use but which use information about the engineer's level of awareness of the functions associated with the controls that are operated to infer situation awareness have also been suggested. Riley (2007, p. 2) provides an example in which "...the engineer sounds the horn when approaching grade crossings, maintenance areas, and so forth. This suggests that a smart cognitive alerter could infer that the engineer is aware of the location of the train when using the horn." FRA's efforts to help develop cognitive alerters or other fitness for duty devices were largely unsuccessful. Railroads balked at the cost of replacing and maintaining new devices which proved effective, and many of the other devices failed rigorous testing.

Scheduling Tools and Training

As was noted previously in this chapter, the scheduling of work for T&E crews was found by Pollard (1991) and the GAO (1993) to be one of the factors involved in producing fatigue in these employees. The principle of "first in, first out" does not take into consideration the physiology of sleep and does not maximize scheduled work given the available employees. The consequence is that employees do not sign-on for work due to fatigue and personal schedules, resulting in a shortfall in employees available to do the required work. This results in more fatigue and stress to the available employees because of excessive overtime and utilization of staff on scheduled off-duty days. Fatigue and stress are the result of uncertainty in the start time of the next job, long working hours, commutes and waiting times.

Across industries, "…the vast majority of supervisors and managers who perform shiftwork scheduling and management have no formal training in the human factors of sleep, the health and safety impacts of shiftwork, or the array of methods and formulas for designing and managing low stress-high productivity work schedules and shift rotations…" (Federal Railroad Administration, 2011c, p. 2). On-the-job training and experience are often the only sources of instruction in Work Schedule Management (WSM) (Gertler and Nash, 2004a).

The Rail Fatigue Management and Evaluation Model in Table 7.2 recognizes scheduling as an important element of the FRA fatigue program strategy. Implementing that strategy, however, was not articulated beyond recognizing that validated biomathematical fatigue models such as FAST and FAID could be used as scheduling tools. The question was, who was going to use such tools at the railroads to do WSM?

Mike Coplen developed tactics to encourage railroads to train personnel to become certified WSMs. The training would involve the use of scientific principles derived from sleep research and assisted by tools designed specifically to optimize work schedules for available workers given consideration of Hours of Service rules, schedule type, consecutive work hours and days, days off and their equitable distribution, shift start and stop times, quality and quantity of weekend time-off, and off duty time between consecutive workdays (Gertler and Nash, 2004a,b). Other tools include procedures for conducting a staffing analysis and inventory, for visualizing, analyzing and evaluating work schedules, and a checklist for evaluating scheduling tools. These and other tools are conveniently described and illustrated in Gertler, Popkin, Nelson & O'Neil (2002). FAST and FAID were not validated and calibrated at that time.

Given the railroad industry's reliance on informal on-the-job training, a project was launched to determine what the formal training needs of WSMs would be. This was done, using state of the art scientific methods called a "strategic job analysis" (Morrow et al., 2010; Federal Railroad Administration, 2011c). The Morrow et al. study examined the job tasks WSMs and documented the knowledge, skills, abilities, and other characteristics (KSAOs) needed to perform the job of a WSM. To accomplish this, they interviewed and surveyed subject matter experts in academia and consulting, and WSM job incumbents in organizations across a wide range of industries. They determined that additional training was required concerning the effects of work schedules on worker health, safety and performance. Moreover, WSMs needed information about the design of ergonomic schedules as recommended by BEST practices (Wedderburn, 1991), and the ability to evaluate the scheduling system that they use to improve the system. Finally, the report recommended that critical gaps in WSM job tasks and KSAOs be remedied by the development of a certification program for WSMs.

Other reports demonstrated the value of conducting staffing analyses and using scheduling tools to reduce fatigue and improve the quality of work life by suggesting alternative schedules that provided job coverage with the same number of employees (Gertler and Nash, 2004a,b).

Positive Train Control (PTC)

Table 7.2 also notes the potential of PTC to ameliorate the effects of fatigue in the railroad industry. This technology is designed specifically to address four issues: "PTC systems are required to prevent train-to-train collisions, over-speed derailments, incursions into work zone limits, and the movement of a train through a switch left in the wrong position" (United States Government Accountability Office, 2019, footnote 7). The concern is specifically with locomotive crew fatigue which can cause all of the types of accidents that PTC is designed to prevent. This is a technology that is designed to prevent human error by taking control of the train from the person operating the train if certain conditions are met. For instance, a fatigued locomotive engineer might have a lapse or microsleep when passing an advance signal and, consequently, not be able to stop the train before passing a stop signal. Signals passed at danger (SPADs) can result in train-to-train collisions. PTC would detect the speed of the train as it traveled between the advance signal and the stop signal and apply emergency braking to prevent the SPAD. At the time of this writing, PTC has yet to be implemented broadly in the railroad industry as witnessed by the referenced GAO report above. Consequently, it remains to be seen if fatigue-caused accidents will be affected by this technology, and a number of human factors concerns remain to be resolved. It is almost axiomatic in the human factors field that technological solutions to human errors are often accompanied by new, unanticipated human errors.

Ergonomics

Chapter 8 will deal with the broader aspects of human factors considerations regarding ergonomics in the railroad industry. Table 7.2 lists environmental influences as factors that affect fatigue. In the locomotive cab such influences include temperature, noise and vibration. However, FRA did not fund projects to assess the influence on these environmental on fatigue because of the difficulty of performing such studies. While there are numerous anecdotal reports about the interaction of fatigue with temperature, noise and vibration, there are few rigorous studies to substantiate the anecdotes (Kryter, 1985; Brooke and Ellis, 1992; Griffin, 1992; Hygge, 1992; Smith and Jones, 1992). Environmental conditions are known to affect performance, as does fatigue, but it remains to be demonstrated whether these conditions combine independently (additively) or synergistically (e.g., multiplicatively).

Railway Worker Fatigue and Organizational Culture

Again, Table 7.2 lists "Understand and influence the organizational work culture" as a need/policy issue for the FRA fatigue program. Educational programs are cited in Table 7.2 as one of the elements of a fatigue program, and the Fatigue website has already been discussed above in this chapter under *Railway worker fatigue and technology*. Three other safety culture projects also address fatigue, but are much broader in scope. These include the Confidential close call reporting system (C3RS), Peer-to-peer observations, and Accident investigations. A full discussion of these projects can be found in Chapter 9 on Safety Culture.

NOTES

1 Commonly abbreviated U.S.C.
2 Commonly abbreviated 49 CFR 228.
3 Commonly abbreviated Pub.L.
4 Commonly abbreviated Stat.
5 Included in this group was the two major labor unions that represent conductors and engineers (United Transportation Union (UTU) and the Brotherhood of Locomotive Engineers (BLE), and the Vice Presidents of CSX Transportation, Conrail, Burlington Northern Railroad, Illinois Central Railroad, and the Union Pacific Railroad.
6 The AAR is the industry trade group that represents the Class I railroads in the U.S.
7 This is partly the reason why FRA did not succeed in having changes made in the Hours of Services laws when reauthorization of the agency was being considered in Congress until the passage of RSIA.
8 Railroads are classified by the Surface Transportation Board according to annual operating revenue. Due to inflation, this criterion changes periodically. Class I railroads have the highest operating revenue. In 2020, that criterion was $900 million. Included in Class I railroads are CSX Transportation, Canadian National Railway, Norfolk Southern, BNSF, Canadian Pacific Railway, Kansas City Southern Railway, Union Pacific Railroad, and Amtrak. Class II railroads have the second highest operation revenue at $40.4 million. Class II railroads are often regional railroads such as the Florida East Coast Railway. Class III railroads have operating revenue of less than $40.4 million and are usually short line railroads (AAR, 2021).
9 This refers to an unbroken period of sleep. A person may have one or more sleep periods in a day (e.g., a long period of sleep at night and several short duration naps). The duration of all those periods would constitute the total daily sleep, while the average sleep period would be the arithmetic mean of those sleep periods. Thus the average sleep period would be less than the total daily sleep.
10 The Human Factors Coordinating Committee was established in 1991 by the Secretary of Transportation to become the focal point for human factors issues within DOT. The Committee is a multi-modal team (air, rail, highway, marine transportation) with government-wide liaisons (Office of the Secretary of Transportation, the Research and Innovative Technology Administration, The Volpe National Transportation Systems Center, the National Transportation Safety Board, the National Institute for Occupational Safety and Health, the Department of Homeland Security, the US Coast Guard, the Transportation Security Administration, the Department of Defense, and the Department of Commerce). It has successfully addressed crosscutting human factors issues in transportation. For more information see https://hfcc.dot.gov/index.html.
11 On the first day the engineers were familiarized with the route they would operate over on the four experimental days.
12 The alerter is a device which is intended to determine if the locomotive engineer is alert and functioning. Operating the throttle, brakes, horn or bell serves to prevent the alerter from putting the train into a sequence that results in emergency braking. In a typical alerter, a light will flash after a period of time that is inversely related to the train's speed. If a button on the alerter is not pushed, or one of the above operations is performed, in a period of time, an audible alarm is sounded. After another period of time has elapsed, if an action has not been taken to reset the alerter, the train is put into emergency braking.
13 Validity in this context means that a fatigue model predicts changes in job performance and/or job-related errors, such as incidents and accidents, caused by fatigue.

14 The measure. cognitive effectiveness, tracks speed of performance on a simple reaction time test and is strongly related to overall response speed, vigilance, and the probability of lapses (Hursh, et al., 2004; Van Dongen, 2004). Cognitive effectiveness can have values in the range of 0–100. A person who consistently obtains 8 hr of good quality sleep would have a peak effectiveness score of 100 during the following waking period.

15 This can be thought of as analogous to the temperature at which water boils (100°C).

16 Actigraphy is a validated method for assessing sleep durations and sleep/wake activity in healthy adults (Ancoli-Israel et al., 2003). Actigraph devices are the size of a wrist watch and must be worn at all times to produce a valid record of sleep periods and durations.

17 200,000 employee hours (e–h) is a standard exposure metric for occupational accidents. This is the e–h for 100 employees working 40 hours/week for 50 weeks.

REFERENCES

Ahlbom, A. (1993). *Biostatistics for epidemiologists*. Boca Raton, FL: Lewis.

Ancoli-Israel, S., Cole, R., Alessi, C., Chambers, M., Moorcroft, W., & Pollack, C. (2003). The role of actrigraphy in the study of sleep and circadian rhythms. *Sleep, 26*(3), 342–392.

Arnedt, J.T., Wilde, G.J., Munt, P.W., & MacLean, A.W. (2001). How do prolonged wakefulness and alcohol compare in the decrements they produce on a simulated driving task? *Accident Analysis and Prevention, 33*, 3, 337–44.

Association of American Railroads. (1993, September). *Work/Rest review task force. A railroad labor and management effort to investigate and address on-the-job-fatigue.* Paper presented at the Federal Railroad Administration Human Factors Research Meeting. Washington, DC.

Association of American Railroads. (1994, November). *Letter to the board with attachment on the status of the Work/Rest review task force.* Washington, DC: Association of American Railroads.

Association of American Railroads. (2021). *Railroad facts* (2021 edn). Washington, DC: Association of American Railroads.

Balkin, T. J., Braun, A. R., & Wesensten, N. J. (2002). The process of awakening: a PET study of regional brain activity patterns mediating the reestablishment of alertness and consciousness. *Brain, 125*, 2308–2319.

Balkin, T. J, Thorne, D., Sing, H. et al. (2000). *Effects of sleep schedules on commercial driver performance* (Report No. DOT-MC-00-133). Washington, DC: U.S. Department of Transportation, Federal Motor Carrier Safety Administration.

Berg, B. L. (1995). *Qualitative research methods for the social sciences* (2nd edn). Boston: Allyn and Bacon.

Bonnet, M. H. (1997). Sleep fragmentation as the cause of daytime sleepiness and reduced performance. *Wien Med Wochenschr, 146*, 332–334.

Brooke, S., & Ellis, H. (1992). Cold. In D. M. Jones & A. P. Smith (Eds.), *Handbook of human performance: The physical environment* (Vol. 1, pp. 105–130). London: Academic Press.

Calabrese, C., Mejia, B., McInnis, C., France, M., Nadler, E., & Raslear, T.G. (2017). Time of day effects on railroad roadway worker injury risk. *Journal of Safety Research, 61*, 53–64. https://dx.doi.org/10.1016/j.jsr.2017.02.007

Carskadon, M., & Dement, W. (1977). Sleep tendency: An objective measure of sleep loss. *Sleep Research, 6*, 200.

Centre for Sleep Research. (2003). *Fitness for duty*. Unpublished manuscript.

Coplen, M., & Sussman, E. D. (2000). Fatigue and alertness in the United States railroad industry Part II: Fatigue research in the Office of Research and Development at the Federal Railroad Administration. *Transportation Research Part F, 3*, 221–228.

Craig, A., & Cooper, R. E. (1992). Symptoms of acute and chronic fatigue. In A. P. Smith & D. M. Jones (Eds.), *Handbook of human performance* (Vol. 3, pp. 289–339). New York: Academic Press.

Devoe, D. B., & Abernethy, C. N. (1977). *Maintaining alertness in railroad locomotive crews* (Report No. DOT/FRA/ORD-77/22). Washington, DC: U.S. Department of Transportation.

Dinges, D. F., Orne, M., & Orne, E. (1985). Sleep depth and other factors associated with performance upon abrupt awakening. *Sleep Research, 14,* 92.

Dinges, D. F., & Powell, J. W. (1985). Microcomputer analysis of performance on a portable, simple visual RT task during sustained operations. *Behavioral Research Methods, Instrumentation, and Computers, 17,* 652–665.

Dinges, D. F., & Powell, J. W. (1989). Sleepiness impairs optimum response capability. *Sleep Research, 18,* 366.

Evans, M., Hastings, N., & Peacock, B. (2000). *Statistical distributions* (3rd edn). New York: Wiley.

Falmagne, J.-C. (1985). *Elements of psychophysical theory.* New York: Oxford University Press.

Federal Motor Carrier Safety Administration and Federal Railroad Administration (2016, March). *Advance notice of rulemaking on obstructive sleep apnea* (Docket No. FRA-2015-0111). Washington, DC: Department of Transportation. Accessed at https://www.fra.dot.gov/eLib/details/L17364#p1_z5_gD_ksleep%20apnea.

Federal Motor Carrier Safety Administration and Federal Railroad Administration. (2017). Evaluation of safety sensitive personnel for moderate-to-severe obstructive sleep apnea. *Federal Register, 82,* 37038–37039. https://www.govinfo.gov/content/pkg/FR-2017-08-08/pdf/2017-16451.pdf.

Federal Railroad Administration. (1998). *Charter of the North American rail alertness partnership.* doi:10.13140/RG.2.2.17413.60644

Federal Railroad Administration. (2004). Notice of safety advisory 2004-04; effect of sleep disorders on safety of railroad operations. *Federal Register, 69,* 58995–58996. https://search.usa.gov/search?query=safety+advisory+2004-04&op=GO&affiliate=fra

Federal Railroad Administration. (2008). *Work schedules and sleep pattern survey data.* https://www.fra.dot.gov/Page/P0604.

Federal Railroad Administration. (2011a). *Measurement and estimation of sleep in railroad worker employees* (RR11-02). https://www.fra.dot.gov/rpd/downloads/RR_Measurement_and_estimation_of_Sleep_final_1102.pdf.

Federal Railroad Administration. (2011b). *The fatigue accident validation database.* https://www.fra.dot.gov/eLib/details/L03039

Federal Railroad Administration. (2011c). *Training and certification of work schedule managers may improve shift scheduling practices* (Report No. RR11-03). Washington, DC: U.S. Department of Transportation.

Folkard, S., & Akerstedt, T. (1987). Towards a model for the prediction of alertness and/or fatigue on different sleep/wake schedules. In A. Oginiski, J. Pokorski, & J. Rutenfranz (Eds.), *Contemporary advances in shiftwork research.* Krakow: Medical Academy. pp. 231–240.

Folkard, S., & Akerstedt, T. (1992). A three process model of the regulation of alertness-sleepiness. In R. Oglivie & R. Broughton (Eds.), *Sleep, arousal and performance problems and promises. A tribute to Bob Wilkinson* (pp. 11–26). Boston: Birkhauser.

Froberg, J. (1977). Twenty-four-hour patterns in human performance, subjective and physiological variables and differences between morning and evening active subjects. *Biological Psychology, 5,* 119–134.

Gamst, F. C. (1980). *The hoghead. An industrial ethnography of the locomotive engineer.* New York: Holt, Rinehart and Winston.

Gertler, J., & DiFiore, A. (2009). *Work schedules and sleep patterns of railroad train and engine service workers* (Report No. DOT/FRA/ORD-09/22). Washington, DC: U.S. Department of Transportation.

Gertler, J., & DiFiore, A. (2011). *Work schedules and sleep patterns of railroad train and engine service employees in passenger operations* (Report No. DOT/FRA/ORD-11/5). Washington, DC: U.S. Department of Transportation.

Gertler, J., DiFiore, A., & Raslear, T. (2013). *Fatigue Status of the U.S. Railroad Industry.* (Report No. DOT/FRA/ORD-13/06). Washington, DC: U.S. Department of Transportation.

Gertler, J., Hursh, S. R., Fanzone, J., & Raslear, T. G. (2012). *Validation of FAST model sleep estimates with actigraph measured sleep in locomotive engineers* (Report No. DOT/FRA/ORD-12/05). Washington, DC: U.S. Department of Transportation.

Gertler, J., & Nash, D. (2004a). *Optimizing staffing levels and schedules for railroad dispatching centers* (Report No. DOT/FRA/ORD-04/01). Washington, DC: U.S. Department of Transportation.

Gertler, J., & Nash, D. (2004b). *Reducing railroad dispatcher fatigue with alternative work schedules* (Paper No. TRB2004-001697). Paper presented at the 84th Annual Meeting of the Transportation Research Board, Washington, DC.

Gertler, J., Popkin, S., Nelson, D., & O'Neil, K. (2002). *Toolbox for transit operator fatigue* (TCRP Report 81). Washington, DC: National Academy Press.

Gertler, J., & Viale, A. (2006a). *Work schedules and sleep patterns of railroad signalmen* (Report No. DOT/FRA/ORD-06/19). Washington, DC: U.S. Department of Transportation.

Gertler, J., & Viale, A. (2006b). *Work schedules and sleep patterns of railroad maintenance of way workers* (Report No. DOT/FRA/ORD-06/25). Washington, DC: U.S. Department of Transportation.

Gertler, J., & Viale, A. (2007). *Work schedules and sleep patterns of railroad dispatchers* (Report No. DOT/FRA/ORD-07/11). Washington, DC: U.S. Department of Transportation.

Griffin, M. J. (1992). Vibration. In D. M. Jones and A. P. Smith (Eds.), *Handbook of human performance: The physical environment* (Vol. 1, pp. 55–78). London: Academic Press.

Halcrow Pacific. (2006). *International search & review of engineering safety devices utilised on rail systems* (Doc No 63600). Sidney Australia: Halcrow Pacific.

Harrison, Y., & Horne, J. A. (1996). Long-term extension to sleep—are we really chronically sleep deprived? *Psychophysiology, 33,* 22–30.

Hays, W. L. (1963). *Statistics for psychologists.* New York: Holt, Rinehart and Winston.

Hursh S. R., Balkin, T. J., Miller, J. C., & Eddy, D. R. (2004). The fatigue avoidance scheduling tool: Modeling to minimize the effects of fatigue on cognitive performance. *SAE Transactions, 113,* 1, 111–119.

Hursh, S., Fanzone, J., & Raslear, T. (2011). *Analysis of the relationship between operator effectiveness measures and economic impacts of rail accidents.* (DOT/FRA/ORD-11/13). Washington, DC: Federal Railroad Administration. (Available at https://www.fra.dot.gov/eLib/details/L01301)

Hursh, S. R., Raslear, T. G., Kaye, A. S., & Fanzone, J. F. (2006). *Validation and Calibration of a Fatigue Assessment Tool for Railroad Work Schedules, Summary Report* (Report No. DOT/FRA/ORD-06/21). Washington, DC: U.S. Department of Transportation.

Hursh, S. R., Raslear, T. G., Kaye, A. S., & Fanzone, J. F. (2008). *Validation and Calibration of a Fatigue Assessment Tool for Railroad Work Schedules, Final Report* (Report No. DOT/FRA/ORD-08/04). Washington, DC: U.S. Department of Transportation.

Hursh S. R., Redmond, D. P., Johnson, M. L., Thorne, D. R., Belenky, G., Balkin, T. J, Storm, W. F., Miller, J. C., & Eddy, D. R. (2004). Fatigue models for applied research in warfighting. *Aviation, Space and Environmental Medicine, 75,* 3, Suppl.: A44–53.

Human Factors Coordinating Committee. (2003). *U.S. DOT operator fatigue management program. Partnerships for 21st Century transportation safety and security.* Washington, DC: Human Factors Coordinating Committee.

Hygge, S. (1992). Heat and performance. In D. M. Jones & A. P. Smith (Eds.), *Handbook of human performance: The physical environment* (Vol. 1, pp. 79–104). London, UK: Academic Press.

Jewett, M. (1997). *Models of circadian and homeostatic regulation of human performance and alertness.* [Dissertation]. Cambridge, MA: Harvard University.

Jewett, M., & Kronauer, R. (1999). Interactive mathematical models of subjective alertness and cognitive throughput in humans. *Journal of Biological Rhythms, 4,* 588–597.

Kerlinger, F.N. (1973). *Foundations of behavioral research* (2nd edn). New York: Holt, Rinehart and Winston.

Kryter, K. D. (1985). *The effects of noise on man* (2nd edn). Orlando, FL: Academic Press.

Labin, S. N. (2005). *Macro evaluation model for research programs on human fatigue.* (Department of Transportation Contract No. DTRS57-03-F-50078). Philadelphia, PA: Institute for Survey Research, Temple University. U.S

Lilienfeld, A. M., & Lilienfeld, D. E. (1980). *Foundations of epidemiology* (2nd edn). New York: Oxford University Press.

Luce, R.D. (1986). *Response times.* New York: Oxford University Press.

Luce, R. D., Krantz, D. H., Suppes, P., & Tversky, A. (1990). *Foundations of measurement: Representation, axiomatization, and invariance* (Vol. 3). San Diego, CA: Academic Press.

Lumley, M., Roehrs, T., and Zorick, F. et al. (1986). The alerting effects of naps in sleep-deprived subjects. *Psychophysiology, 23,* 403–408.

Madill, A. (2012). Interviews and interviewing techniques. In H. Cooper, P. M. Camic, D. L. Long, A. T. Panter, D. Rindskopf, & K. J. Sher (Eds.), *APA handbook of research methods in Psychology:* Foundations, *planning, measures, and psychometrics* (Vol. 1). Washington, DC: American Psychological Association.

Morgan, D. L. (1988). *Focus groups as qualitative research.* Newbury Park, CA: Sage.

Mitler, M., Gujavarty, S., Sampson, G., & Bowman, C. (1992). Multiple daytime nap approaches to evaluating the sleepy patient. *Sleep, 5,* 119–127.

Monk, T. H. (1989). A visual analogue scale technique to measure global vigor and affect. *Psychiatry Research, 27,* 89–99.

Monk, T. (1991). Circadian apects of circadian sleepiness: a behavioral messenger? In T. Monk (Ed.), *Sleep, sleepiness and performance* (pp. 39–63). Chichester: Wiley. .

Monk, T., & Embry, D. (1981). A field study of circadian rhythms in actual and interpolated task performance. In A. Reinberg, N. Vieux, & P. Andlauer (Eds.), *Night and shift work: Biological and social aspects* (pp. 473–480). Oxford: Pergamon Press.

Morrow, S., Walsh, B., Badiee, M., Stentz, T., Nash, D., Clark, V. P., Barnes-Farrell, J., & Impara, J. (2010). *Work schedule manager gap analysis: assessing the future training needs of work schedule managers using a strategic job analysis approach* (Report No. DOT/FRA/ORD-10/05). Washington, DC: U.S. Department of Transportation. https://www.fra.dot.gov/eLib/details/L04330#p1_z5_gD_kMorrow

National Transportation Safety Board. (1999). *Evaluation of U.S. Department of Transportation efforts in the 1990s to address operator fatigue* (Safety Report NTSB/SR-99/01). Washington, DC: National Transportation Safety Board.

National Transportation Safety Board. (2018). *Nearly identical probable causes for 2 commuter rail accidents drive safety recommendations.* Washington, DC: National Transportation Safety Board. https://www.ntsb.gov/news/press-releases/Pages/MR20180206.aspx

Neri, D.F. (2004). Fatigue and performance modeling workshop, June 13–14, 2002. *Aviation, Space, and Environmental Medicine, 75,* A1–A3.

Pilcher, J. J., & Coplen, M. K. (2000). Work/rest cycles in railroad operations: effects of shorter than 24-h shift work schedules and on-call schedules on sleep. *Ergonomics, 43,* 573–588.

Pollard, J. K. (1991). *Issues in locomotive crew management and scheduling* (Report No. DOT-VNTSC-FRA-91-1). Washington, DC: U.S. Department of Transportation.

Pollard, J. K. (1996). *Locomotive engineers' activity diaries* (Report No. DOT-VNTSC-FRA-96-12). Washington, DC: U.S. Department of Transportation.

Raslear, T. G. (1993, September). *Unpublished memo on meeting on hours of work.*

Raslear, T. G. (2013a). *Fatigue comparison, pre-RSIA to post-RSIA.* Unpublished PowerPoint presentation.

Raslear, T. G. (2013b). *Criteria and procedures for validating biomathematical models of human performance and fatigue: Procedures for analysis of work schedules* (Report No. DOT/FRA/ORD-13/33; Docket No. FRA-2009–0043-0003). Washington, DC: U.S. Department of Transportation. https://www.fra.dot.gov/eLib/details/L04703#p2_z5_gD_kraslear and https://www.regulations.gov/#!documentDetail;D=FRA-2009-0043-0003

Raslear, T. G. (2014a, February). *Fatigue status of the U.S. railroad industry.* Presentation at the *Federal Railroad Administrator's Fatigue Summit*, February 13, 2014. Washington, DC: Raslear. doi:10.13140/RG.2.2.15551.05284

Raslear, T. G. (2014b). *Start time variability and predictability in railroad train and engine freight and passenger service employees* (Report No. DOT/FRA/ORD-14/05). Washington, DC: U.S. Department of Transportation.

Raslear, T. G. (2014c). Prevalence and treatment of sleep apnea in safety-critical railroad employees. *Journal of Sleep Disorders & Therapy, 3,* 179. doi:10.4172/2167–0277.1000179

Raslear, T. G. (2019). *The GAO reports on hours of service, 1992–1993.* Silver Spring, MD: Raslear. doi:10.13140/RG.2.2.20503.62882.

Raslear, T. G., Gertler, J., & DiFiore, A. (2013): Work schedules, sleep, fatigue, and accidents in the US railroad industry. *Fatigue: Biomedicine, Health & Behavior, 1,* 99–115.

Raslear, T. G., Hursh, S. R., & Van Dongen, H. P. A. (2011). Predicting cognitive impairment and accident risk. In H. P. A. Van Dongen and G. A. Kerkhof (Eds.), *Progress in brain research: Human Sleep and Cognition. Part II: Clinical and Applied Research* (Vol. 190, pp. 155–167). Amsterdam: Elsevier

Richardson, D., Carskadon, M., & Flagg, W. (1978). Excessive daytime sleepiness in man: multiple sleep latency measurement in narcoleptic and control subjects. *Electroencephalography and Clinical Neurophysiology, 45,* 621–627.

Riley, V. (2005). Proof of concept for a smart cognitive alerter system for locomotives. Unpublished manuscript.

Roach, G. D., Fletcher, A., & Dawson, D. (2004). A model to predict work-related fatigue based on hours of work. *Aviation, Space, and Environmental Medicine, 75,* A61–A69.

Smiley, A. M. (1990). The Hinton train disaster. *Accident Analysis and Prevention, 22,* 443–455.

Sussman, E. D., & Raslear, T. G. (2008). Railroad human factors. In D. A. Boehm-Davis (Ed.), *Reviews of human factors and ergonomics* (Vol. 3); Santa Monica, CA: The Human Factors and Ergonomics Society.

Tabak, B., & Raslear, T. G. (2010). *Procedures for validation and calibration of human fatigue models: The fatigue audit interdyne tool* (Report No. DOT/FRA/ORD-10/14). Washington, DC: U.S. Department of Transportation.

Tepas, D. I., Paley, M. J., & Popkin, S. M. (1997). Work schedules and sustained performance. In G. Salvendy (Ed.), *Handbook of human factors and ergonomics* (2nd edn). New York: Wiley.

Thomas, G. R., Raslear, T. G., & Kuehn, G. I. (1997). *The effects of work schedule on train handling performance and sleep of locomotive engineers: A simulator study* (Report No. DOT/FRA/ORD-97/09). Washington, DC: U.S. Department of Transportation.

Thorne, D. R., Genser, S., Sing, H., & Hegge, F. (1983). Plumbing human performance limits during 72 hours of high task load. In *Proceedings of the 24th DRG seminar on the human as a limiting element in military systems* (pp. 17–40). Toronto: Defense and Civil Institute of Environmental Medicine.

Tilley, A., & Brown, S. (1992). Sleep deprivation. In A. P. Smith & D. M. Jones (Eds.), *Handbook of human performance* (Vol. 3, pp. 238–259). New York: Academic Press.

United States General Accounting Office. (1992). *Railroad safety. Engineer work shift length and schedule variability* (Report No. GAO/RCED-92-133). Washington, DC: United States General Accountability Office.

United States General Accounting Office. (1993). *Railroad safety. Human factor accidents and issues affecting engineer work schedules* (GAO/RCED-93–160BR). Washington, DC: United States General Accountability Office.

United States Government Accountability Office. (2011). *Freight railroad safety. Hours of service changes have increased rest time, but more can be done to address fatigue risks* (GAO 11-853). Washington, DC: United States General Accountability Office.

United States General Accountability Office. (2019). *Positive train control. As implementation progresses, focus turns to the complexities of achieving system interoperability* (Report No. GAO-19-693T). Washington, DC: United States General Accountability Office.

Van Dongen, H. P. A. (2004). Comparison of mathematical model predictions to experimental data of fatigue and performance. *Aviation, Space and Environmental Medicine; 75*(3, Suppl.), A15–36.

Wedderburn, A. (1991). Guidelines for shiftworkers. *Bulletin of European Shiftwork Topics.* Dublin, Ireland: European Foundation for the Improvement of Living and Working Conditions.

Wesensten, N. J., Balkin, T. J., & Belenky, G. (1999). Does sleep fragmentation impact recuperation? A review and reanalysis. *Journal of Sleep Research, 8*, 237–246.

Whitlock, A. (2002). *Driver vigilance devices: systems review* (03 T024 QUIN 22 RPT Final Report Issue 01). United Kingdom: Railway Safety.

Wright, N. A., Stone, B. M., Horberry, T. J., & Reed, N. (2007). *A review of in-vehicle sleepiness detection devices* (Published Project Report PPR 157). United Kingdom: TRL Limited.

8 Working Conditions and Ergonomics

BACKGROUND

Most work in the railroad industry involves both physical and mental activity. While the locomotive crew is operating the train, they are focused on planning the setting of controls to meet the operating conditions that are expected from knowledge of the route that has been memorized from extensive experience. These, clearly, are mental activities. However, mechanical problems, which have a predominately physical aspect, also have to be addressed, such as setting hand brakes on cars and replacing couplers and connections between cars. Many of the ergonomic issues that the FRA Human Factors Research program dealt with during my tenure were centered around Locomotive Cab Working Conditions. Within the Sociotechnical System described in Chapter 1 (see Figure 1.3), this chapter is dealing with the physical system of the locomotive cab and the environment in which it operates as well as the overall physical system of the railroad (e.g., track ballast). It is also concerned with the effect of that physical system on the behavior of individual workers and the teams that they form.

The Department of Transportation (DOT) and FRA were created by the Department of Transportation Act of 1966 (Public Law 89-670). This law became effective in 1967 and authorized DOT to conduct research in unspecified areas related to transportation. The Federal Railroad Safety Act of 1970 was more specific and provided that the Secretary of Transportation (authority delegated to FRA) "…is authorized to perform such acts including, but not limited to, conducting investigations, making reports, issuing subpoenas, requiring production of documents, taking depositions, prescribing recordkeeping and reporting requirements, carrying out and contracting for research, development, testing, evaluation, and training (particularly with respect to those aspects of railroad safety which he finds to be in need of prompt attention), and delegating to any public bodies or qualified persons, functions respecting examination, inspecting, and testing of railroad facilities, equipment, rolling stock, operations, or persons, as he deems necessary to carry out the provisions of this title." (Public Law 91-458, Sec 208 (a)). Importantly, the law appropriated $21,000,000 for Fiscal Years 1971, 1972 and 1973. Of course, not all of that funding was apportioned to research and development.

However, PL 91-458 apparently did get noticed by some members of the academic community. Frederick Gamst published a paper in the journal *Human Factors* in 1975, based on his experience as a railroad employee in train and engine service and from anthropological field research (Gamst, 1975). The paper was focused on the locomotive cab. Among other issues, Gamst called out problems with "Environmental Quality". These included temperature, noise, vibration, air quality (diesel fumes and sand), and sanitary facilities. The Gamst paper was purely qualitative; that is, it did not attempt to

DOI: 10.1201/9781003500018-12

document by measurements the noise level, etc., in the locomotive cab. Consequently, these were all issues that required attention from the regulator, and that involved quantification through impartial research. Fred served on many of the future Railroad Safety Advisory Committee (RSAC) meetings that considered these issues, and his 1975 paper may be considered germinal. It took a while for the FRA to consider the all the issues in Gamst's paper. For instance, in the 1976 Report to Congress FRA reported "Work completed or initiated in 1976 in the human factors area included (1) the award of two contracts to develop alternative deigns for a Research Locomotive and Train Handling Evaluator; (2) completion of a survey of Alcohol and Drug Abuse Programs in the railroad industry; (3) completion of a study of air contaminants present in locomotive cabs; (4) completion of a comparative analysis and evaluation of various locomotive crew alerting devices; (5) completion of the major portions of locomotive cab design development study; (6) initiation of efforts to develop a simplified locomotive in-cab noise measurement procedures; and (7) initiation of a study of personal injury risks associated with working around railroad rolling stock." (FRA, 1976, pp. 47–48).

The study of air contaminants in locomotive cabs found that, even in the longest railroad tunnel in the U.S. (Cascade Tunnel, 7.8 miles), the level of noxious diesel fumes did not exceed current standards set by the Occupational Safety and Health Administration (Hobbs et al., 1977). The same conclusion was reached by Peay and Sanders (1978).

Aurelius (1971, p. 1) reported, in a limited study of locomotive noise levels, that "…The crew's working environment was found to approach the exposure limits set in the Walsh-Healey Public Contracts Act…". The Walsh-Healey Public Contracts Act, Safety and Health Standards of 1969 states that for noise levels above 90 dBA[1] employees should either wear personal protective gear or have time-limited exposure, dependent on the noise level that exceeds the 90 dBA threshold. Aurelius found that noise exposures ranged from 88 dBA (SD-40 locomotive, heavy load, low speed, no horn) to 105 dBA (AGP-20 locomotive, medium load, medium speed, air brake application). A 1-h daily exposure at 105 dBA was permitted under the Act.

In 1977, the Report to Congress (FRA, 1977) indicated that a study had been completed to assess the noise exposure of locomotive crews. There is no Report to Congress for 1978[2] in the FRA eLibrary, and there is no mention of a noise report in the Report to Congress in 1979 (FRA, 1979). Remington and Rudd (1976) did report on locomotive generated noise, but that report mostly focused on noise outside the locomotive. In-cab measurements were reported in Table 4.2 and included a variety of operating conditions including speed (10–55 mph), throttle position (idle to 8), window open or closed, and direction of travel (forward, reverse). They reported noise levels that ranged from 68 to 85.5 dBA. Remington and Rudd did not make measurements with the train horn activated or the air brake in operation, and both of these events are noisy. However, Kilmer (1980) did make measurements with horn and brake and found that "The noise exposure is within acceptable limits because the operational duty cycle is such that the sources which generate high sound levels (horn and brake) are operating only for short periods of time and because the locomotive spends a great deal of time in idle (diesel engine sound levels below 90 dB)." (1980, p. i). This will be revisited when modern standards of noise exposure and the FRA hearing conservation program are discussed below.

Subsequent Reports to Congress on the Federal Railroad Safety Act of 1970 that were available on the FRA website (FRA, 1981, 1982, 1983, 1984) did not include mention of locomotive cab environment issues. However, FRA did publish a noise measurement handbook in 1982 (Stusnick et al., 1982).

THE LOCOMOTIVE CRASHWORTHINESS AND CAB WORKING CONDITIONS REPORT TO CONGRESS

Herewith we focus on the Federal Railroad Administration's Report to Congress (1996) and subsequent efforts involved in the regulation of locomotive cab working conditions. During my first year at FRA (1993) Garold Thomas and I were summoned to the office of Rolf Mowatt-Larsen in the Office of Safety. Rolf had a large flow chart (see Figure 1.1, FRA, Office of Safety Assurance and Compliance, 1996) on his office wall that documented the course of actions that FRA was taking to address cab working condition issues as mandated by Congress in 1992 as a part of the Rail Safety Enforcement and Review Act (RSERA, Public Law 102-365). According to the Federal Railroad Administration's Report to Congress (1996): "In response to the mandate of Section 10 of Public Law 102-365, the Federal Railroad Administration (FRA) prepared a plan of action and milestones for the research and analysis necessary to determine (a) the health and safety effects of locomotive cab working conditions, (b) the effectiveness of Association of American Railroads (AAR) Specification S-580, and (c) the benefits and costs of additional locomotive crashworthiness features. In an effort to fully address the broad range of issues presented in the Act, FRA outlined a multi-faceted approach that included the following:

Conduct of an industry-wide public meeting to gather information from all segments of the industry regarding the areas of concern identified in the Act;

Establishment of a comprehensive locomotive collision data base based on detailed accident information gathered during actual collisions;

Establishment of a research contract to develop and verify a computer model capable of predicting how each of the crashworthiness features in AAR S-580 and in the Act affect the collision dynamics and probability of crew injury; and

Conduct of a detailed survey of the locomotive crew's cab working conditions and environment." (FRA, 1996, p. 1-1).

These tasks were required to be completed in 30 months. Garold Thomas and I were tasked with contracting air quality tests in the Cascade Tunnel (see below, the section on Air Quality).

The FRA Office of Safety was largely responsible for the Report to Congress (FRA, 1996). Locomotive cab working conditions were studied in a 2-year, nationwide investigation that involved more than 230 locomotives from 13 Class I freight railroads and Amtrak. "The study found that locomotive cab conditions need improvement. Considerations of the report included crew hours, cab temperature, cab noise, cab air quality, cab sanitary facilities, cab ergonomics, and cab vibration." (FRA, 1996, p. iii). It is a comprehensive report, but it suffers from several study design flaws, in particular the use of FRA safety inspectors, rather than professional researchers, to conduct measurements and make observations. Many FRA safety inspectors come from the railroad industry, and their judgments concerning cab conditions that they have worked in as railroad employees and observed as an FRA inspector may be influenced by cognitive biases (Kahneman et al., 1982) or by a lack of experience in making scientific measurements. For instance, regarding judgments based on observation, safety inspectors were asked to "Describe the noise generated by the heating and air conditioning system and whether it is conducive to stress even within FRA standards." (FRA, 1996, p. H-2), and, "How Do Cab Environmental Factors Appear to Affect the Ability of the Crew to Safely Operate the Train? Locomotive?" (FRA, 1996, p. H-4, Figure H.1). It is not clear in the report what was done with such subjective judgments. They were not reported in the Summary of Cab Survey Results (FRA, 1996, Appendix I), although the Locomotive Cab Environmental Survey Rating Definitions (p. I-1) describes "Notations from inspection" (presumably responses to questions, such as above) with 5-point Likert scale adjectives: 5, excellent or very good; 4, good; 3, acceptable, fair or average; 2, poor; 1, very bad, very dirty, or unusable. The technical expertise of the FRA inspectors to make these judgments is unclear.

The design of the survey was poor in that it precluded a rigorous statistical analysis. More than 200 observations were made during winter months, but only 30 observations were made during summer months. Although the report states that "... randomly-selected locomotives..." (p. 5-11) were observed, there is no mention of a method for random selection so it must be assumed that a convenience sample was obtained. All of the locomotives were assessed by only one inspector, precluding an analysis of the consistency or reliability of the observations. Moreover, 6% of the cells in the survey data are blank (see Guilford, 1954, p. 289, on dealing with incomplete data matrices). Temperature readings were made, but only reported in the survey data as Likert scale values (1 to 5).

Even within the limits set by design issues, the survey data are under-analyzed. For instance, when using a Likert rating scale, it is desirable to determine several overall characteristics of the data. For instance, there is a tendency for raters to not use extreme categories (i.e., "excellent or very good"; "very bad, very dirty, or unusable"). This is referred to as the "error of central tendency" (Guilford, 1954), and it results in observations being displaced toward the mean of the rating categories (i.e., "acceptable, fair or average"). Figure 8.1 shows the distribution of ratings for the winter and summer observations. It is clear that both distributions, especially the winter distribution, are skewed to the right, away from the mean category towards the more positive categories. The mean rating for winter was 3.53 and for summer was 3.25. The skew in the distributions argues against an error of central tendency, but may indicate a halo effect. If the inspectors, in general, had a positive opinion of locomotive cab conditions prior to the observations, the ratings would be skewed toward the positive ratings. Unfortunately, the design of the survey does not allow a systematic exploration of these two error types (or several others, see Guilford, 1954, pp. 278–288) using statistical tools.

It is also important to perform a test of internal consistency of the ratings. This refers to the degree to which the ratings for an individual item correlate to the total

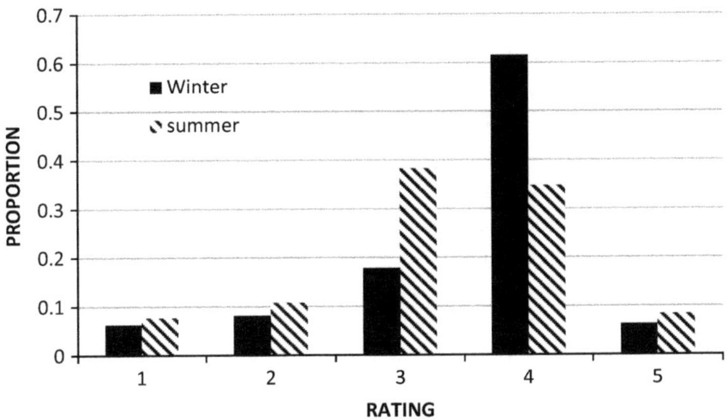

FIGURE 8.1 Rating distributions for winter and summer.

rating score for all items. No numerical guidelines were provided by the sources I read concerning the magnitude a correlation needed to be considered acceptable (e.g., Dunn-Rankin, 1983, p. 92), so I adopted the criterion of statistical significance. That is, I considered the correlation adequate if it was significant at the 0.01 probability level. For winter observations, the only item that failed this test was the condition of the air conditioning system, for which there were very few (42) observations, and should not be considered. The mean correlation for winter was 0.36, which is often considered a middling correlation which would only account for 13% of the variance. For summer, which had far fewer observations, four items did not meet the criterion: toilet vent, passage condition, floor condition, and air valve exhaust. The mean correlation for summer was 0.58 and accounts for 34% of the variance. Overall, the survey appears to have adequate, if moderate, internal consistency.

Finally, any subjective rating scale should be tested for reliability. This refers to the consistency of measurements across different instances of the observation procedure (Joint Committee on the Standards for Educational and Psychological Testing, 2014, p. 33; Strout and Lane, 2012). Reliability can be assessed in a number of ways, including the comparison of rater scores on the same individual (i.e., locomotive). The survey design, however, preempts that simple solution. A common alternative is the use of Cronbach's α statistic (Strout and Lane, 2012). For winter observations Cronbach's α was 0.82, and for summer it was 0.88. Values of Cronbach's α in this range indicate good internal consistency.

Based on this limited analysis of the quality of the 1996 Report to Congress, it can be said that the report provides a starting point for understanding the locomotive cab working conditions at that time. Information that was considered in the report included, amongst other items: current FRA regulations; standards from other industries or government agencies; complaints from railroad employees, labor unions and members of Congress; injuries and illnesses reported to FRA by railroads that were considered relevant to a particular working condition. The findings of the report are briefly summarized below.

TEMPERATURE

The then current FRA regulation, 29 CFR 229.119(d), specifies that "The cab shall be provided with proper ventilation and with a heating arrangement that maintains a temperature of at least 50°F 6 inches above the center of each seat in the cab." (FRA, 1996, p. 5-2).

There were four complaints to the FRA between 1989 and 1993 concerning heat and heaters in the locomotive cab. It is not clear from the report whether these were all the complaints, or just the ones that FRA could verify (FRA, 1996, p. 5-9). There were three casualties in the FRA accident/incident database between 1990 and November 1994 related to cab temperature (FRA, 1996, pp. 5-9–5-10). The report describes standards from several industries and agencies and the effects of temperature on human performance (pp. 5-2–5-8).

Summer temperatures had an average rating of 2.07 (poor), but as the report shows in Figure 5.4 the distribution of scores was heavily skewed toward unfavorable scores with 12/30 scores of "1; very bad, very dirty, or unusable" and 12/30 scores of "2; poor" (p. 5-13). 40% of the cab temperatures were between 80 and 90°F and 40% were above 90°F. Actual temperatures were not provided in the report, so it is not possible to know what the true mean temperature was among the observations. However, the report did note that a high temperature of 121°F was observed.

Winter temperatures had an average rating of 3.51 (acceptable). Figure 5.5 of the report shows that the distribution of scores was skewed toward favorable scores with 93/200 scores of "4; good" and 4/200 scores of "5; excellent". 82% of cab temperatures were found to be above 60°F. Given that FRA regulations at the time only specified a low temperature for the cab, it is not surprising that only 3/200 locomotives had a temperature below 50°F.

NOISE

The then current FRA regulation, 29 CFR 229.121, specifies that "…the permissible exposure to a continuous noise in a locomotive cab shall not exceed an 8-h time-weighted average of 90 dB(A), with a doubling rate of 5 dB(A). Exposure to continuous noise shall not exceed 115 dB(A)…" (FRA, 1996, p. 6-2).

The report describes standards from several industries and agencies (p. 6-3) and the effects of noise on human performance (pp. 6-7–6-9). FRA occupational noise level limits are mostly consistent with those of Occupational Safety and Health Administration (OSHA). Allowable exposure durations for FRA as a function of noise level are shown in Table 8.1. These are generally the same as for (OSHA) with

TABLE 8.1

FRA Occupational Noise Level Limits

Allowable exposure time (h)	12	8	6	4	3	2	1.5	1	0.5	0.25
dBA	87	90	92	95	97	100	102	105	110	115

Source: After FRA (1996, Table 6-1).

the exception of 87 dBA which is included because locomotive crews at the time could work for 12 h.

Between 1989 and 1991, there were no reports of noise-induced injuries or illness. Between 1992 and 1993, however, there were 42 such reports. There were 11 complaints to FRA concerning noise in the 1989–1993 period.

Figure 8.2 is based on Figure 6.3 from the report. It shows the distribution of Average Noise Levels (Lav) and Time-Weighted Average (TWA) noise levels measured in 350 locomotive cabs.[3] The data points are fit with lines that represent a moving average of 2 (see Tukey, 1977 for information about smoothing data). The mean of the Lav distribution is 84.8 dBA with a standard deviation of 3.9 dBA. The median of the Lav distribution is 84.5 dBA, which means that 50% of the measurements are higher than 84.5 dBA. The mean of the TWA distribution is 83.1 dBA with a standard deviation of 4.6 dBA. The median of the TWA distribution is 82.5 dBA. The TWA distribution is skewed toward lower noise levels, a result that is expected since the report notes that the average duration of noise measurements was 6.5 h (TWA is normalized to 8 h; Berger et al., 2000, p. 74). Accordingly, half of the TWA measurements are greater than 82.5 dBA.

The report provides some detail concerning various thresholds (TWAs) established by FRA and OSHA, and the percentage of measurements that exceed those thresholds (see Table 6-4 in the report). These measurements depend on the type of locomotive (lead road, trailing, switching) and crew member (engineer, conductor). Some interesting statistics from that table include: 67% of engineers and 46% of conductors exceed the then current OSHA hearing conservation limit of 85 dBA; 25% of engineers and 10% of conductors exceed the FRA and OSHA 8-h TWA limit of 90 dBA; 38% of engineers and 17% of conductors exceed the FRA 12-h TWA limit of 87 dBA. A contributor to cab noise levels is the air brake vent which, at the time, was located in the locomotive engineer's side of the cab. This may explain the differences seen above between the noise exposure of engineers and conductors.

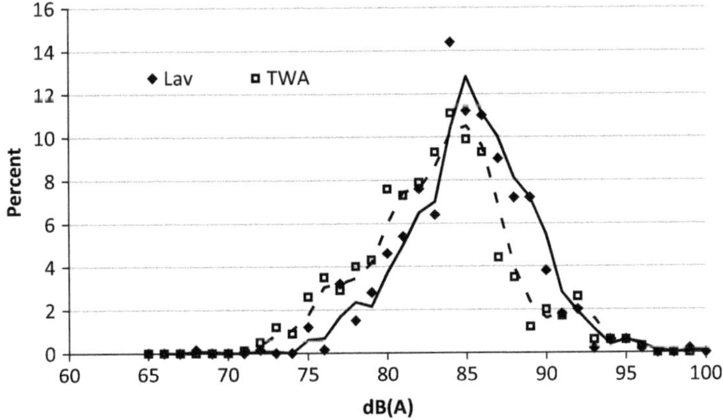

FIGURE 8.2 Distribution of average noise level (Lav) and time-weighted average noise level (TWA) in locomotive cabs. (After FRA, 1996, Figure 6.3).

The report also notes that train horns contribute to the noise levels in the locomotive cab. Train horns are required to produce between 96 and 110 dBA 100 feet in front of the locomotive. One inspector found that with the cab window open the noise level was 106 dBA.

Overall, the 1996 report identifies serious levels of noise exposure to locomotive crews and some of the putative sources of that noise. Also, the report notes that there is no hearing conservation program in FRA regulations.

AIR QUALITY

In 1996 the FRA regulation concerning air quality (49 CFR 229.43) was very limited: "(a) Products of combustion shall be released entirely outside the cab and other compartments. Exhaust stacks shall be of sufficient height or other means provided to prevent entry of products of combustion into the cab or other compartments under usual operating conditions. (b) Battery containers shall be vented and batteries kept from gassing excessively." (FRA, 1996, pp. 7-1–7-2). FRA apparently relied upon OSHA limits (29 CFR 1910.1000) for NO, NO_2, SO_2, CO and CO_2 (Table 7.1) to enforce the Locomotive Inspection Act of 1994 (PL103-272). The report also notes military and Federal Aviation Administration standards concerning air quality. Although there is a section that claims to discuss the effects on human performance of exposure to these fumes, the discussion is almost entirely focused on health effects.

The report notes that there were 34 complaints concerning fumes in the locomotive cab between 1989 and 1993. Between 1990 and 1994 there were 46–60 reports to FRA of injuries or illnesses due to fumes in the locomotive cab. The complaints and reported illnesses often involved operations through tunnels, and frequently concerned eye and throat irritation.

Consequently, FRA decided to conduct air quality tests in the Cascade Tunnel. As noted above two studies in the 1970s of air quality in the locomotive cab found that levels of measured contaminants were within OSHA standards at that time. The Cascade Tunnel was chosen because "...(1) it is the longest railroad tunnel in the U.S. — longest enclosure available; (2) there is evidence that trains overheat due to idling in the tunnel — overheating leads to dieseling which produces numerous gases; and (3) diesel engines must strain to traverse the length of the tunnel." (FRA, 1996, p. 7-8). Measurements were conducted in April and December, 1994. Hydrocarbons (including benzene), CO, CO_2, NO, NO_2, SO_2, O_3, particulates, and stibine were measured. Twenty-eight locomotives, some in trailing positions, were tested. Six lead locomotives were tested in April by an industrial hygiene contractor who found that all measurements were below OSHA exposure limits. A more extensive set of measurements were made in December on 22 locomotives by FRA inspectors. In December measurements were made at 2, 10, and 20 minutes into the tunnel, yielding a total of 198 samples of NO, NO_2 and CO. Table 7-5 of the report shows that two measurements exceeded permissible exposure levels (PEL) for NO, and one measurement exceeded PEL for NO_2. A total of 13 measurements were made for O_3, CO_2, SO_2, CH_3CHO, and hydrocarbons. None exceeded PELs. The mean rating for all gases in Table 7-5 is 4.77 (on a 1–5 scale based on measured concentrations).

The mean "subjective" rating from Appendix I (Cab Survey), on the same scale, is 4.37 for the winter observations and 4.24 for the summer.

These results are consistent with measurements made in the 1970s. However, the 1996 report does not provide a reference to a formal report on the April or December 1994 tests. Consequently, many details of the testing procedure are unknown, and the raw data (concentrations of the gases measured) are not provided, only a 1–5 Likert scale. Considering the cost and effort taken to conduct these tests, some public record of the procedures and test data should have been preserved to resolve issues consequent to 1996. For instance, the 1996 report states that NO was 50 ppm in locomotive #12 at 10 min, and was 25 ppm at 20 min. This is not consistent with the rating scale noted in Table 7-5 (rating = 1), which Table 7-4 indicates signifies that the concentration of NO was at least 35 ppm.

SANITATION

"FRA does not have regulations that require sanitary facilities, toilets or food storage (i.e., coolers or refrigerators) on locomotives." in 1996 (FRA, 1996, p. 8-2). At the time, FRA relied on the Locomotive Inspection Law of 1968 for its ability to inspect and enforce guidelines for sanitation in locomotives. Such guidelines were provided by OSHA's rules on workplace sanitation in 29 CFR 1910.141. However, states could opt out of OSHA rules by applying their own, and, consequently, enforcement of OSHA rules depended on the location of the locomotive.

To make matters worse, in 1996 "The FRA accident/incident data base is constructed such that any current or past injury or illness due to a faulty sanitary condition cannot be specifically identified, and consequently, contains no reports from railroads of employee injuries or illnesses directly attributable to locomotive cab sanitary conditions." (FRA, 1996, p. 8-3). The report also does not cite any employee or labor union complaints concerning sanitary conditions as it does for other cab working condition issues.

In the locomotive cab survey, in which over 200 cabs were observed, seven items concerned sanitation including the condition of the toilet, the water cooler, food storage, toilet vent, passage condition, floor condition, and the cleanliness of the cab. Again, these subjective observations were rated on a 1–5 scale. These ratings are not statistically different between summer and winter observations, and the grand means are shown in Table 8.2.

While these ratings are all close to the "acceptable" category (3), the 1996 report notes that "Thirty percent (58 of 197) of the locomotives equipped with toilets were rated as unacceptable or poor, and 30% (60 of 197) had toilet spaces rated to have

---

TABLE 8.2

Mean Ratings for Sanitary Conditions from the 1996 Locomotive Cab Survey

	Toilet Condition	Water Cooler	Food Storage	Toilet Vent	Passage Condition	Floor Condition	Cab Cleanliness
Rating	2.94	3.4	3.37	2.67	3.59	3.67	3.38

unacceptable or poor ventilation. Thirty-two percent (75 of 234) of the locomotives were not equipped with water coolers, 27% (60 of 234) were not equipped with food storage facilities, and 16% (37 of 234) had cab interiors that were judged to be inadequate for general sanitation." (FRA, 1996, pp. 8-6–8-7)

The report also documents the types of toilets in use (port-a-let, Microphor, and dry hopper[4]) and the types of chemical agents that are used in the toilets to disinfect them. The 1996 report concludes, among other things, that sanitation is an occupational health threat to the locomotive crew (p. 8-10–8-11).

CAB ERGONOMICS

The 1996 report states that "Current FRA regulations pertaining to cab layout and design features are very limited and provide minimal guidance with respect to specific design features to be incorporated in the locomotive cab." (FRA, 1996, p. 9-1). The report references 49 CFR 229.119 which mentions cab seats, windows, doors, and floors, but does not specify design features. The report recommends standardization and redundancy for controls and displays, and appears to rely on pre-publication Human Factors guidelines that were prepared for the FRA Office of Research and Development and published in 1998 (Multer et al., 1998).

Between 1989 and 1993, there were 21 complaints about cab ergonomics. Eight concerned the seats, four concerned slip/trip hazards, three concerned door latches, two concerned drinking water and two concerned gauge lights. The remaining two were complaints about garbage and obstructions. Between 1990 and November 1994, there was an average of 281 casualties reported to the FRA due to cab layout (FRA, 1996, Table 9-2), and an average of 63 casualties due to cab seats (FRA, 1996, Table 9-3). There are 14 cause codes for general cab layout and four for cab seats. The ratio of layout casualties to seat casualties is 4.5, and the ratio of cause codes is 3.5. This suggests that the higher number of casualties for cab layout might be due to the larger number of cause codes for cab layout (see Chapter 3, Data Issues).

The typical control panel in 1996 was the AAR's locomotive control stand, shown in Figure 8.3 along with other aspects of the locomotive cab. The control stand contains various meters to monitor speed, the status of the air brake system,

FIGURE 8.3   Locomotive cab showing the AAR control stand.

and the status of the diesel engine. Controls include the throttle, the reverser (to change direction of travel), the dynamic brake, the independent brake, the train brake, the train horn and the bell. This locomotive is equipped with windshield wipers, window shades, a water cooler, food storage, and seats with backs and arm rests. The 1996 report recommends that all locomotive cabs have these "amenities", noting that many cabs do not. The control stand is designed so that the locomotive can be operated in either direction by the engineer by turning his seat around.

VIBRATION

In 1996, vibration was not regulated by the FRA. The report discussed sources of vibration (engine, suspension, track) and the transmission of vibration to the crew through seats that are fixed to the floor of the cab. Standards were being developed by the International Organization for Standardization (ISO), and the U.S. military had standards in place to mitigate issues of health and safety, work efficiency, and comfort. The report discusses some of the then known effects of whole-body vibration and some of the possible ways to control vibration in the locomotive cab. No measurements of vibration were made for the report, and the survey did not include subjective observations of vibration.

THE DEVELOPMENT OF REGULATIONS AFTER
THE 1996 REPORT

The staff of the Office of Research and Development in the Division of Equipment, Operations, and Hazardous Materials was highly involved in the development of regulations concerned with working conditions and ergonomics following the 1996 report. Operating Practices was the then, current appellation for Human Factors (FRA, Office of Research and Development, 1992). Little original research was required as there was an established literature on many of the topics that would be addressed. Literature searches for the pertinent and creditable information was needed, as was the application of that information to the railroad environment and operations. As one of the very few staff at FRA who had knowledge and training in human behavior, physiology, sensation and perception, and cognition, I became the person who regularly attended meetings of Working Groups and Task Forces created by the Railroad Safety Advisory Committee (RSAC, see Chapter 1, p. 11, Chapter 2, p. 21 on the regulatory process) to gather information and propose regulations that would satisfy the requirements of FRA, railroad management, railroad manufacturers, and railroad labor to improve working conditions and safety while allowing the efficient operation of the railroad system. When information in the literature was inadequate or difficult to apply to railroad operations and the railroad environment, government contracts were awarded to conduct original research to answer specific questions concerning temperature, noise, sanitation and vibration in the locomotive cab. Also, the biomechanics of working in the larger railroad environment (e.g., walking on ballast, operating hand brakes) came to be considered. Finally, issues concerning the standardization of signage in the railroad industry were also researched.

TEMPERATURE

As noted above, regulations concerning locomotive cab temperature only specified that that 50°F was the lowest compliant temperature. Since no upper limit was specified in 29 CFR 229.119(d), there was no regulatory rationale for systematic measurements of high temperatures by the FRA's compliance officials, the inspectors in the Office of Safety. Consequently, there was no data available to the FRA concerning high temperatures when temperature was discussed in RSAC working and task groups. I am a firm believer in the adage that you can't manage what you don't measure, so a primary task for the Human Factors staff was getting measurements of temperature in the locomotive cab during cold and hot weather. A plan for doing this was devised by the Human Factors staff, and Office of Safety inspectors were detailed to conduct the measurements on several types of locomotives. Unfortunately, only the cold weather measurements were obtained. These are summarized in Raslear (1998/2020a). The memo states "The locomotive cab temperatures, on the average, appear to have been maintained within the limits desired. A temperature of 55°F was the lowest acceptable cab temperature if the external temperature was >0°F. The average temperature minus the standard deviation, or the trimmed mean minus the standard deviation, are all above 55°F in Tables 1–5. On his basis, it seems unlikely that many temperatures were encountered that were unacceptable." (p. 1).

However, a review of the, then, currently available literature indicated that temperature extremes were a health and safety problem for workers. For example, Ramsey, Burford, Beshir and Jensen (1983) showed that unsafe behaviors at an industrial plant approximated a quadratic function of ambient temperature. Unsafe behaviors were at a minimum at a particular temperature, but increased as ambient temperature deviated from that temperature. Raslear (1998/2020b) saw two specific problems in the Ramsey et al. report. First, unsafe behaviors are not accidents; second, industrial plants are not the same as the railroad environment. Consequently, it was requested that the FRA examine its database for accidents to determine if a relationship exists between the occurrence of railroad accidents and ambient temperature. Figure 8.4 shows the outcome of that request.

The relationship shown in Figure 8.4 is based on accident statistics obtained from the Federal Railroad Administration's database for the years 1992–1997. The cause categories are all from the general category, "Operating Practices". Cause categories

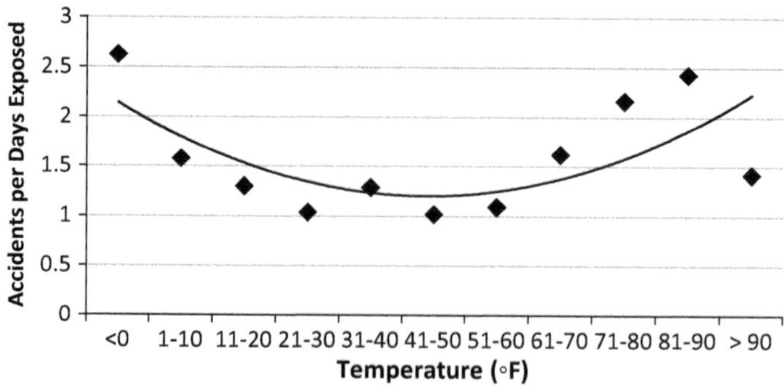

FIGURE 8.4 Accidents per day exposed as a function of temperature.

that would involve outdoor operations, and therefore exposure to climatic extremes, were selected from the list of all possible operating practices causes so that the ambient temperature in which the accidents occurred would also be the temperature to which the employee was likely exposed. Raslear (1998/2020b, Table 1) lists the number of accidents by cause and temperature range.

Figure 8.4 shows accidents per day exposed as a function of ambient temperature. The metric, accidents per day exposed, is used because temperature "extremes" do not occur as frequently as temperatures within a "normal" range. Consequently, we expect that there are fewer days with "extreme" temperatures within a year. Because of this, there are fewer opportunities to observe accidents at "extreme" temperatures, and a plot of accident frequency as a function of temperature would show a low number of accidents at "extreme" temperatures and a majority of accidents at temperatures in a "normal" range. Thus, a simple plot of accident frequency as a function of temperature confounds the effect of temperature on accident occurrence with the unequal frequency of all temperatures. The relationship between accident occurrence and temperature can be easily unconfounded by dividing accident frequency in a temperature range by the number of days that temperatures in that range occur. The result is the metric, accidents per day exposed, used in Figure 8.4.

Temperature in the continental United States (i.e., excluding Hawaii) has an approximately normal distribution based on historical data obtained from the National Weather Service web page (https://www.ncei.noaa.gov/access/monitoring/climate-at-a-glance/national/time-series/110/tavg/12/12/1895-2022?base_prd=true&firstbaseyear=1901&lastbaseyear=2000). Temperature data has a mean of 51.5 °F and a standard deviation of 20 °F. This means that in a year, the mean temperature for the U.S. will be between 31.5 and 71.5°F 68% of the time, or for 248 days. Knowing the mean and standard deviation of the temperature distribution allows the expected number of days per year in any temperature range to be calculated. The procedure is to find the z-score[5] corresponding to the upper and lower bounds of the range, and to then find the probabilities[6] corresponding to those z-scores. Those probabilities are subtracted to determine the probability of the temperature range. The number of days in the temperature range is 365 times the number of years of data (6) times the probability of the temperature range.

A quadratic equation was fit to the data in Figure 8.4, as well as a linear equation. For the quadratic fit, the correlation coefficient was 0.77 ($p < 0.01$), indicating that a curvilinear relationship, similar to that seen in the Ramsey et al. study, is statistically reliable and accounts for more than 59% of the variability in the data. For the linear fit, the correlation coefficient was 0.15 ($p > 0.05$, not statistically reliable) and accounts for only 2.3% of the variance.

On the basis of the information contained in Raslear (1998/2020b), it was concluded that there is an approximately quadratic relationship between railroad accidents and ambient temperature (see also the analysis of Reinach, and Gertler (2001, Section 4.2.5) who obtained a similar result.)

To further the rationale for a regulation that included hot temperature limits, the Human Factors staff authorized a meta-analysis[7] (Hunter and Schmidt, 1990) of the effects of temperature of human performance. This resulted in a peer-reviewed journal article by Pilcher et al. (2002). An overview of this study was provided by Raslear (2003) in a PowerPoint presentation to the RSAC working on the temperature regulation.

Pilcher et al. identified 527 articles, reports and dissertations published between 1922 and 1997 of which 226 were primary studies of temperature effects on performance. Criteria for inclusion in the analysis required:

> hot or cold environmental exposure as experimental condition;
> hot exposure could be quantified as Wet Bulb Globe Temperature (WBGT);
> cold exposure included air temperature;
> studies with exposure by water, clothing, head gear were excluded;
> neutral temperature ranges within defined limits:
>> hot: 60–69.6°F WGBT
>> cold: 65–75;
> each included study reported at least one performance measure:
>> reaction time
>> attention/perceptual
>> mathematical processing
>> reasoning, learning, memory;
> studies using only motor tasks, self-reports and physiological measures were
>> excluded
> effect size was capable of computation
> mean and standard deviation.

These criteria were met by 23 of 226 primary studies and were used in the meta-analysis.

The analysis examined the following variables:

type of temperature exposure
Hot (>70°F WBGT)

Hot1 (70–79.9°F WBGT)
Hot2 (80–89.9°F WBGT)
Hot3 (>90°F WBGT)

Cold (<65°F)

Cold1 (50–64.9°F)
Cold2 (<50°F)

Duration of experimental session

Short (<120 min)
Long (>120 min)

Duration of pre-task exposure

none
short (1–59 min)
long (>60 min)

Type of performance task

    reaction time
    attention/perceptual
    mathematical processing
    reasoning, learning, memory

Duration of task battery

    short (<60 min)
    long (>60 min).

Figure 8.5 shows the mean percent difference in performance relative to the neutral temperature groups for the five temperature subcategories described above. The means were calculated across all conditions within each temperature subcategory. Figure 8.5 has been plotted so as to emphasize the similarity of Figure 8.5 with Figure 8.4 and with the findings of Ramsey et al. (1983) that unsafe behaviors and accidents are a quadratic function of temperature. Hot and cold temperatures cause decrements in performance.

Figure 8.6 shows performance decrements for each of the four categories of task as a function of temperature. The abscissa in Figure 8.6 is plotted normally. Figure 8.6 shows that temperatures have different effects on the performance of each type of task. Learning, memory and reasoning tasks are highly affected by cold temperatures, while math, perception and attention are highly affected by hot temperatures. The performances affected are all components of tasks performed by locomotive engineers and conductors in their jobs. In the context of accidents, performance decrements are called human errors, unsafe or at-risk behaviors.

Accidents have multiple causes, including human errors (Reason's "swiss cheese" model, Reason, 1990, 1997). Various factors, including temperature, cause human errors, and, by definition, in Human Factor accidents there is at least one human error. Heinrich (1959) recognized that there are many more errors or unsafe behaviors than

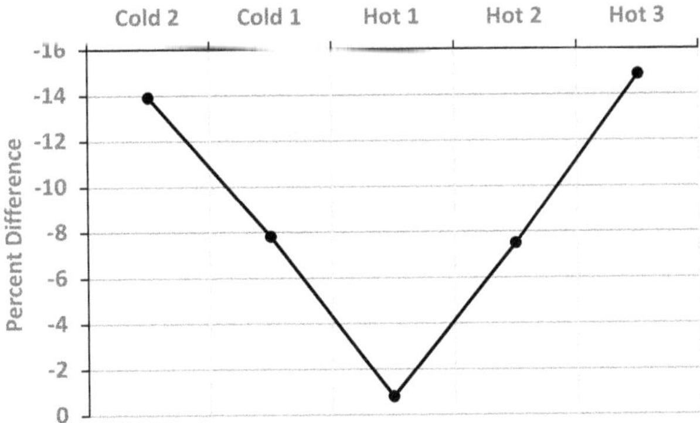

FIGURE 8.5   The mean percent difference in performance between neutral temperature groups and the five temperature subcategories. (From Pilcher et al., 2002, Table 2).

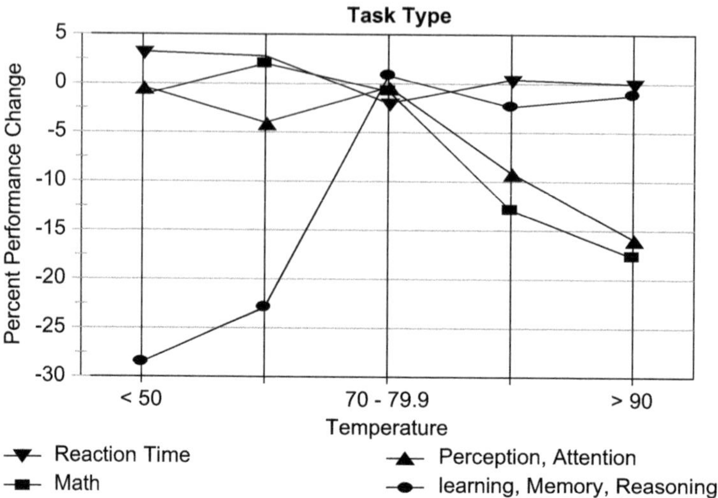

FIGURE 8.6   Percent performance decrement for four types of tasks as a function of temperature subcategory. (Data from Pilcher et al. 2002.)

accidents. This is often expressed in what is commonly called Heinrich's accident triangle (Figure 8.7).

Although the ratios presented in Figure 8.7 are disputed, the idea that there are many more fatal and serious accidents than minor injuries and unsafe behaviors or errors is commonly accepted. Consequently, we can say that Human Factors accidents (A) are proportional to errors (E):

$$A \propto E, \tag{8.1}$$

therefore, there is some constant of proportionality, b, for which

$$A = bE. \tag{8.2}$$

Since temperature causes an increase in human error, controlling temperature in the cab working environment can reduce errors and accidents by c

$$cA = cbE, \tag{8.3}$$

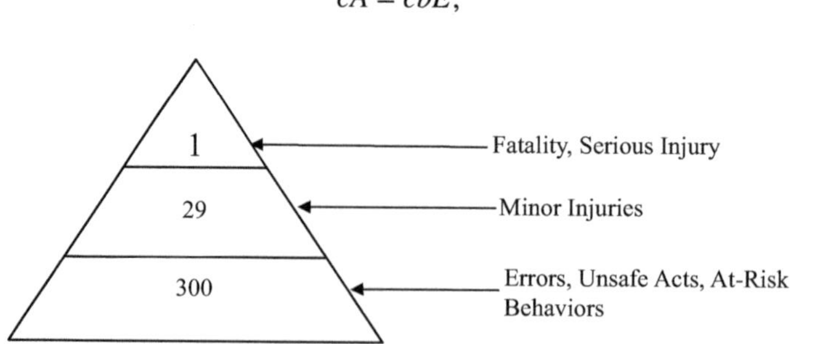

FIGURE 8.7   Heinrich's accident triangle.

where c is the percentage performance effect (Figure 8.5) expressed as a proportion. This means that by keeping locomotive cab temperatures within the range, $50°F \leq T \leq 90°F$, approximately 15% of human factor accidents could be avoided (cab temperatures are already mandated to be above 50°F).

In 2012 FRA revised the regulations regarding locomotive cab temperature (Federal Railroad Administration, 2012). The new regulations cite Pilcher et al. (2002) to support an increase in the minimum temperature in the cab from 50 to 60°F, but do not set a maximum temperature: "FRA believes that the issues need to be considered further before a determination can be made as to whether a maximum temperature requirement would be appropriate." (p. 21320). Rather, the regulation mandates air conditioning in the cab which must be maintained in proper operating condition. The Cold 1 condition in Pilcher et al. covers the temperature range $50°F < T \leq 64.9°F$. For that subcategory, the reduction in performance was 7.81%. A 10° change over a 15° range means that one could expect, approximately, a 5.23% decrease in human factor accidents since air conditioners are only required by the regulation for new locomotives, effective June 8, 2012. In 2011, there were 738 human factor train accidents, and in 2014 there were 692. This is a 6% reduction in human factor train accidents.

The Multer et al. (1998) report made specific recommendations for temperature in the locomotive cab based on the Human Factors literature. The report recommends a minimum temperature of 64°F and a maximum of 85°F. The temperature gradient from floor level to head level should not vary more than 10°F. Glazing, insulation and humidity recommendations, as well as others, were also made in the report.

Noise

FRA proposed Task 97-2 (locomotive cab working conditions) to the RSAC in 1997. A Working Group was formed by consent of the RSAC to "...(1) Revise existing cab noise limits to take into account current requirements of the OSHA standard, specifically as it relates to hearing conservation programs, and (2) Continue efforts to evaluate engineering controls and other measures used to minimize noise exposure in locomotive cabs." (Federal Railroad Administration, 2006, p. 63070). The OSHA standard, which can be found at 29 CFR 1910.95, has a permissible 8-h time weighted average (TWA) exposure level of 90 dB(A), a 5 dB exchange rate,[8] and a 8-h TWA 90 dB(A) threshold.[9] The OSHA rules also require a hearing conservation program (HCP) at regulated entities. The final rule which resulted from Task 97-2 was published in 2006 and was effective in 2007.

It took FRA 10 years to formulate a rule on noise exposure in the locomotive cab because the topic is highly complicated and technical. For instance, there are several sources of noise in the cab, some of which have safety implications outside of, or in addition to occupants of the locomotive cab. Train horns are required to warn motorists and others of the train's approach at grade crossings. Horns are required to have a sound level of between 96 and 110 dB at 100 feet in front of the locomotive. These sound levels are very high in the cab and contribute to noise exposure for the crew, and also produce community noise annoyance. Reducing horn sound levels, however, increases the risk of collisions at grade crossings which is a safety concern for the crew and motorists. Engine noise, bells, radios, air conditioning, brakes, and loose components all produce noise in the locomotive cab. Discussions concerning engineering and administrative controls for these noise sources were exhaustive.

The amended noise regulations, 49 CFR 227 and 49 CFR 228, are described and discussed in Federal Railroad Administration (2006). A summary of the major provisions of the regulations are that: "FRA is amending its occupational noise standards for railroad employees whose predominant noise exposure occurs in the locomotive cab. FRA's previous standard (issued in 1980) limited cab employee noise exposure to certain levels based on the duration of their exposure. This rule modifies that standard and also sets out additional requirements.

FRA is requiring railroads to conduct noise monitoring and to implement a hearing conservation program for railroad operating employees whose noise exposure equals or exceeds an 8-h time-weighted average (TWA) of 85 decibels. FRA is also establishing design, build, and maintenance standards for new locomotives and maintenance requirements for existing locomotives." (Federal Railroad Administration, 2006, p. 63066).

The amended rule requires railroads to limit noise exposure of locomotive cab employees to an 8-h TWA of 90 dB(A). As noted above, this is consistent with the OSHA noise rule. (The Multer et al. (1998) report recommends a maximum continuous noise level of 75 dBA as a goal.) However, FRA requires that employees who are exposed to an 8-h TWA of 85 dB(A) or higher to be enrolled in a hearing conservation program. The OSHA rule sets that threshold at 90 dB(A). The FRA exchange rate is set at 5 dB, again consistent with the OSHA rule.

Numerous hours were spent discussing the most appropriate exchange rate for the railroad industry. While OSHA and the Mine Safety and Health Administration (MSHA) uses a 5 dB rate, the Air Force and the National Institute of Occupational Safety and Health uses a 3 dB rate. A 3 dB exchange rate represents a doubling of sound energy, consistent with the hypothesis that hearing damage is due to sound energy. FRA settled on a 5 dB rate for considerations of feasibility and cost (Federal Railroad Administration, 2006, p. 63071). However, there may be some confusion about what a decibel refers to in the definition, and, consequently, what the true impact is of a particular exchange rate.

With regard to sound, decibels (dB) can refer to either measurements of sound power (watts, W) or to sound pressure (pascals, Pa). Sound power is proportional to the square of sound pressure (W $\propto$ Pa^2). From the rule (49 FR 227.5, Federal Railroad Administration, 2006, p. 63125): "*Sound level or Sound pressure level* means ten times the common logarithm of the ratio of the square of the measured A-weighted sound pressure to the square of the standard reference pressure of twenty micropascals, measured in decibels."

Mathematically, for sound pressure:

$$\text{Sound pressure level (SPL)} = 10\log\frac{\text{Pa}^2}{\text{Pa}_0^2}\,\text{re } 20\ \mu\text{Pa}\,, \tag{8.4}$$

$$\text{Sound pressure level (SPL)} = 20\log\frac{\text{Pa}}{\text{Pa}_0}\,\text{re } 20\,\mu\text{Pa}. \tag{8.5}$$

For sound power:

$$\text{Sound power level (PWL)} = 10\log\left(\frac{W}{W_0}\right)\text{re } 10^{-16}\text{ W/cm}^2. \qquad (8.6)$$

The reference levels, Pa_0 and W_0 are equivalent measures of the human absolute threshold for hearing. In other words, 20 µPa $= 10^{-16}$W/cm². In Equation (8.6), a doubling of sound power (or energy) results in a 3 dB increase in PWL. However, in Equation (8.5), a doubling of sound pressure results in a 6 dB increase in SPL. Consequently, an exchange rate of 5 dB in SPL is equivalent to a 2.5 dB exchange in PWL. A 5 dB SPL exchange rate is more conservative if energy is the putative cause of hearing damage, but less conservative than a 3 dB exchange rate (equivalent to 1.5 dB PWL). The role of sound energy in hearing damage is not clear at this time. A thorough discussion of this issue can be found in Federal Railroad Administration (2006, pp. 63076–63078).

The FRA rule also allows a brief exposure to very loud noise: "Section 227.105(d) addresses continuous noise exposure above 115 dB(A). This requirement differs from OSHA's standards. OSHA prohibits unprotected exposures above 115 dB(A) (See 29 CFR 1910.95(a) (https://www.ecfr.gov/current/title-29/section-1910.95#p-1910.95(a)) and 29 CFR 1926.52(a) (https://www.ecfr.gov/current/title-29/section-1926.52#p-1926.52(a))). By contrast, FRA permits very brief exposures to continuous noise which is defined as noise that exceeds one second) between 115 dB(A) and 120 dB(A) as long as the total daily duration does not exceed 5 seconds." (Federal Railroad Administration, 2006, pp. 63090–63091).

It should be noted that the FRA Human Factors participant in this RSAC objected to this deviation from OSHA standards because discomfort is reported from exposure to sounds above 115 dB (Licklider, 1951, Fig. 5, p. 995). This is likely an underestimate of the threshold for auditory discomfort because the subjects were first acclimated to intense auditory stimulation (p. 997). It should also be noted that temporary threshold shifts (TTS) of 10 dB occur from exposures to sounds of 3 sec duration at 105 dB (Thurlow, 1971, p. 235). TTS magnitude increases rapidly with increases in SPL (Lindsey, 1981, Fig. 2.11.D.1) and can result in permanent threshold shifts of more than 40 dB after 10 years of daily exposure in the workplace to noise above 100 dB (Lindsey, 1981, Fig. 2.09.C). It is clear that there is a complex relationship between TTS and permanent damage to the auditory system which is dependent on the duration and SPL of the noise (see Kryter, 1985, pp. 219–330). Discomfort, pain and hearing loss should not be acceptable conditions of employment.

The rule places primary responsibility for compliance with the railroads. The rule requires that railroads "Develop and implement a noise monitoring program; administer a hearing conservation program; establish and maintain an audiometric testing program; make audiometric testing available to employees; implement noise operational controls (if desired); require the use of hearing protection; make hearing protection available to employees at no cost; train employees in the use and care of hearing protection; ensure proper fitting of and supervise the correct use of hearing protection; give employees the opportunity to select hearing protection from a variety of suitable hearing protection; evaluate hearing protection attenuation; initiate and offer a training program, maintain and retain records; and obtain and maintain locomotives that meet specified standards for limiting in-cab noise". (Federal Railroad Administration, 2006, p. 63072).

The rule also places responsibilities on the employees. Employees are required to "Use their hearing protection when mandated by the railroad; care for their hearing protection as trained by the railroad; and complete the training program which is offered by the railroad. There is one additional obligation for which employees have primary responsibility—employees must report for audiometric testing once every three years. While railroads have an affirmative obligation to offer testing, employees have an affirmative obligation to report for testing." (Federal Railroad Administration, 2006, pp. 63072–63073).

The requirement for audiometric testing was a particularly contentious issue for railroad labor. Anecdotally, many conductors and locomotive engineers develop hearing loses and sue the railroads for negligence under the provisions of the Federal Employers' Liability Act of 1908 (Committee for Study of the Federal Employers' Liability Act, 1994; also see Chapter 2, p. 19). In the railroad industry this is considered the equivalent of a "golden parachute" for train and engine (T&E) employees because the railroads will frequently settle such suits rather than go through the expense of litigation and public exposure regarding noise conditions in the industry. Railroad labor was concerned that rank-and-file union members would object to the loss of this potential payout from the railroads and were also concerned that the detection of a significant hearing loss would result in their disqualification from T&E employment.

SANITATION

As noted above, FRA did not have regulations concerning sanitary conditions in the locomotive cab prior to 2002. In 1997, FRA proposed that RSAC undertake an examination of the issues involved including toilet systems, washing facilities, potable water, ventilation, lighting, trash disposal, provisions for toilet paper and bottled water, servicing, and unique operations or characteristics that might require specialized regulatory treatment (Federal Railroad Administration, 2002). Consequently, FRA published amendments to 49 CFR 229, effective June 3, 2002, that set "…standards that address toilet and washing facilities for employees who work in locomotive cabs. This rule provides exceptions for certain existing equipment and operations, and establishes servicing requirements." (p. 16032).

The rule requires that all lead locomotives have a toilet (sanitation compartment) that is adequately ventilated and equipped with a door with a modesty lock that closes. The compartment should be equipped with a washing system, sufficient toilet paper, and a trash receptacle. A daily inspection of lead locomotives is required under the rule with penalties for violations. The Multer et al. (1998) report notes several other design issues that might improve the sanitation conditions.

VIBRATION AND BIOMECHANICS

As of August 18, 2022 there are no final rules or notices of proposed rules for vibration in the locomotive cab, or for the biomechanics of work in the railroad industry.

There were no rules regarding vibration prior to 1996. Whole-body vibration is a known health risk (see Griffin, 1990, Chapter 5), and we discuss this issue first.

T&E employees work in an active motion environment. Many of them also experience health problems and are involved in safety related incidents. Are T&E employees at higher risk for health and safety problems as a result of the train motion environment (vibration in particular) than groups that do not work in that environment? The answer to this question requires an epidemiological approach that uses objective data on the locomotive motion environment, data on the health and safety characteristics of populations representative of railroad crew, and data that allows the characterization of the locomotive environment in terms of non-motion factors that may influence health and safety.

Information on the measurement of the effects of mechanical shock and vibration on comfort, well-being, health, and safety is captured and codified in several international standards such as those by the International Organization for Standardization (ISO). Two ISO standards are of particular pertinence for our discussion: ISO 2631-1 (1997) and ISO 9996 (1996).

ISO 2631-1 provides information on the evaluation of the effects of shock and whole body vibration with regard to the measurement and evaluation of vibration, the impact of vibration on health, motion sickness, and the impact of vibration on comfort and perception. Whole body vibration is characterized by its frequency, amplitude, waveform, repetition rate and direction. Frequency (f) is measured in hertz (Hz). The amplitude of a vibration is measured as an acceleration (a, in meters/second^2 (m/s^2)). The direction of the acceleration occurs in 3-dimensional space and has both linear and rotational aspects. Rotational acceleration is measured in radians per second^2 (rad/s^2). Linear accelerations include motion on the x-axis (along the direction of travel), the y-axis (at right angles to the direction of travel), and the z-axis (upwards/downwards perpendicular to the floor). Rotations about the y-axis are termed *pitch*, about the x-axis are termed *roll*, and about the z-axis are termed *yaw*. See Figure 8.8 and Griffin (1990, 1992, 1997).

The frequency range of concern is 0.5–80 Hz for health, comfort and perception, and 0.1–0.5 Hz for motion sickness. "The body is most sensitive to vertical vibration in the 2–20 Hz range, especially the 4–10 Hz range. Up to 2 Hz, the body acts at a

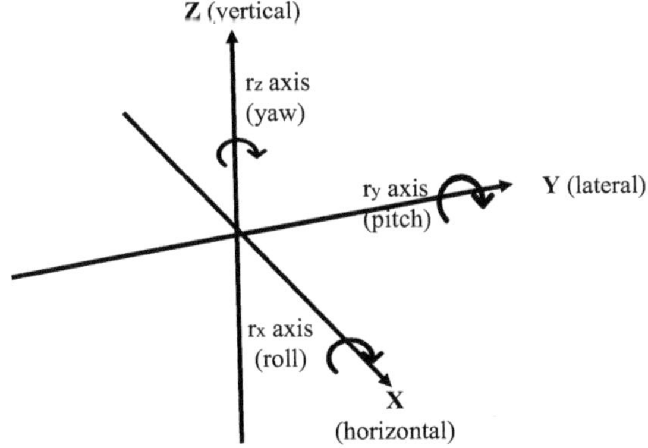

FIGURE 8.8 The six modes of vibrational motion. See text for details.

single mass." (Multer et al., 1988, p. 58). "Biodynamic research as well as epidemiological studies have given evidence for an elevated risk of health impairment due to long-term exposure with high-intensity whole-body vibration. Mainly the lumbar spine and the connected nervous system may be affected. Metabolic and other factors originating from within may have an additional effect on the degeneration. It is sometimes assumed that environmental factors such as body posture, low temperature, and draught can contribute to muscle pain. However, it is unknown if these factors can contribute to the degeneration of discs and vertebrae." (ISO, 1997, p. 21).

However, Griffin (1997, p. 841) notes that "Other disorders which may have been claimed to be due to occupational exposures to whole-body vibration include abdominal pain, digestive disorders, urinary frequency, prostatitis, haemorrhoids, balance and visual disorders, headaches, and sleeplessness."

The effects of whole-body vibration are due to the combination of vibration magnitude, frequency, and duration of exposure. The absolute threshold for the perception of vertical whole body vibration is about 0.001 ms^{-2} rms (root mean squared). An acceleration of 0.1 ms^{-2} rms is noticeable, an acceleration of 1 ms^{-2} rms is uncomfortable, and an acceleration of 10 ms^{-2} rms is dangerous (Griffin, 1997). The ISO 2631 exposure limits for acceleration were set at approximately half the pain threshold for healthy human subjects according to Griffin. The dependency of vibration discomfort on magnitude and duration requires the use of a dose measure for determining exposure limits. The vibration dose value (VDV) that is used in ISO 2631 and British Standard 6841 (1987) is 15 ms$^{-1.75}$. Figure 8.9 shows the dependence of exposure limits on the duration and magnitude of acceleration for a 1 min and an 8 h exposure as a function of vibration frequency. The limit for exposure is lower in general for higher exposure durations, and higher for higher frequencies. Since most occupational vibration exposures would involve a complex mixture of frequencies, it may be more useful to see the exposure limit simply as a function of duration in Figure 8.10.

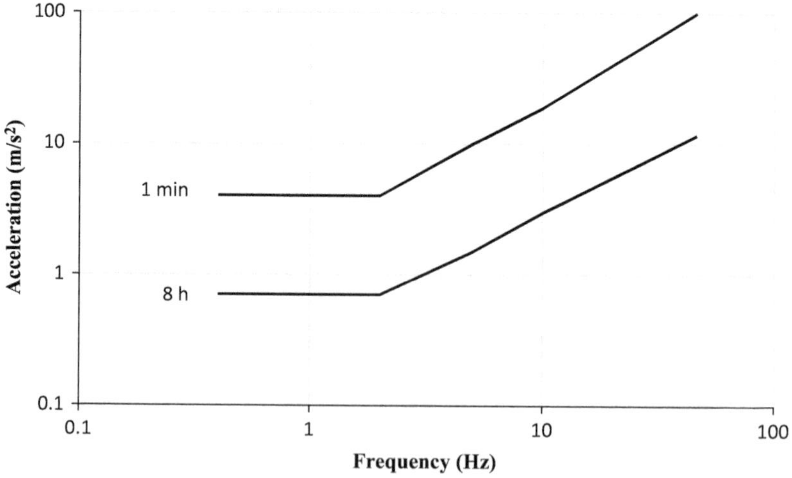

FIGURE 8.9 Exposure limits based on a vibration dose value of 15 ms$^{-1.75}$ (British Standard 6841, 1987, for the *x*- and *y*-axes).

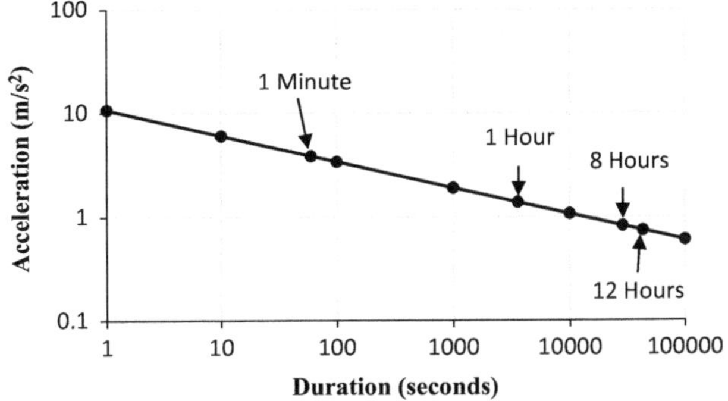

FIGURE 8.10   Acceleration exposure limits as a function of duration, based on an action level of 15 ms$^{-1.75}$ (British Standard 6841, 1987).

Figure 8.11 shows the VDV values for Health Risk and Health Caution zones for vibration exposure durations on linear-linear axes. An excellent, detailed discussion of measurement issues and the effects of duration and frequency on human response to vibration can be found in Griffin (1990).

There is a dearth of data on the epidemiology of the health effects of whole-body vibration for railroad train crews, data on the locomotive motion environment, and data on the health and safety characteristics of populations representative of railroad crew.[10] In general, it should be noted that "There are not sufficient data to show

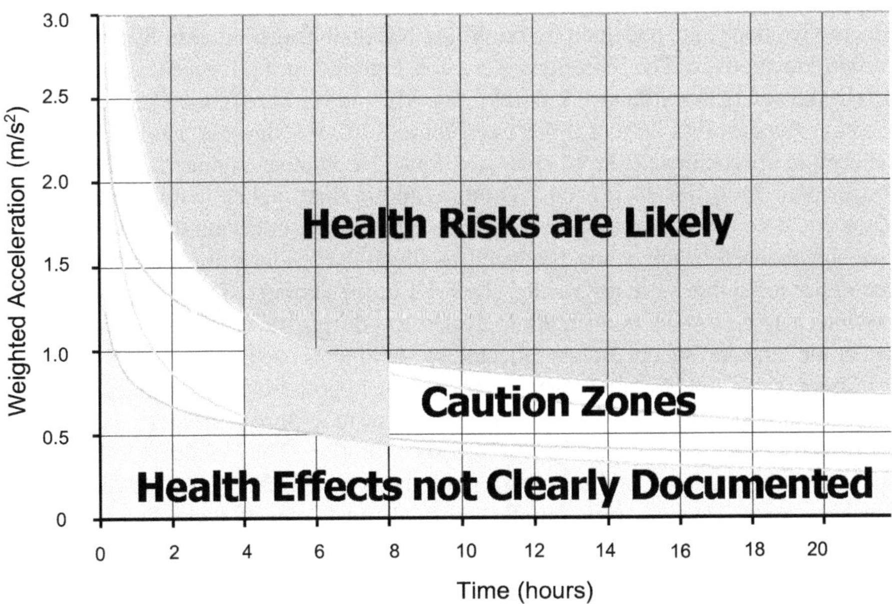

FIGURE 8.11   Weighted acceleration as a function of exposure duration. The curves show the VDV values for health risk and caution zones for vibration exposure. (From Marquis, 2009.)

a quantitative relationship between vibration exposure and risk of health effects. Hence, it is not possible to assess WBV in terms of the probability of risk at various magnitudes and durations." (Marquis, 2009, p. 19). Consequently, the ISO guidance for health and comfort should not be viewed as definitive. An additional problem is that, in his review of the literature, Marquis (p. 36) found "The methods used are disputed by researchers." These cautions are also alluded to by Griffin (1990, Chapters 2 and 3). Consequently, it is not surprising that FRA was unable to formulate a rule that was acceptable to the concerned parties to the RSAC. The following research was available at the time.

Johanning et al. (2006) measured vibration exposure in a randomly selected group of railroad engineers and a sedentary control group. They also surveyed the participants regarding neck and lower back disorders and the comfort of their seats. They found that "...almost all crest factors (CF), MTVV[11] and VDV values were above the critical ratios given in ISO 2631-1." for the railroad engineers. Relative to the sedentary controls, the prevalence of neck and lower back disorders was nearly double in railroad engineers. On a 4-point Likert scale (1 = excellent, 4 = unacceptable) engineers rated their seats poorly (3.02–3.51) in contrast to the controls (1.96–3.44). Johanning et al. found that "Existing cab and seat design in locomotives can result in prolonged forced awkward spinal posture of the operator combined with WBV exposure. In a logistic regression analysis, time at work being bothered by vibration (h/day) was significantly associated with an increased risk of low back pain, shoulder and neck pain, and sciatic pain among railroad engineers."

Smith et al. (2006) used a 6 degree-of-freedom motion (see Figure 8.8) simulator to generate vibration in four types of locomotive seats. The vibrations were generated from locomotive floor data collected by the Volpe National Transportation Systems Center in two locomotives. The vibrations were concentrated in the frequency range below 10 Hz and had peaks around 1.5–2.0 Hz. The VTV lower limit of the Health Guidance Caution Zone in ISO 2631 is 0.743 (see Figure 8.12). For the seat pan this value was reached in approximately 3 h of exposure. Since locomotive engineers typically work longer than 8 h shifts (Gertler and DiFiore, 2009), this implies that the Health Risk Zone could be reached in a typical workday. The authors conclude that "The locomotive signals used in this study produced relatively large and complicated motions in the upper torso that were physically observed in the subjects. These large multi-axis motions may be a major contributor to discomfort during the operation of locomotives under the more severe conditions reflected by the signals used in this study."

Cooperrider and Gordon (2008) examined the hypothesis that isolated shocks and impacts in the locomotive vibration environment may pose a greater health risk to locomotive crews than ordinary whole body vibration. They evaluated shocks and impacts using the VDV and spinal stress methods of ISO 2631 on more than 90 h of measurements. Their analysis found that shocks and impacts had a low probability of adverse health effects.[12]

Johanning et al. (2002) measured the whole body vibration exposure of locomotive engineers and the attenuation of vibration of seats in 22 locomotives. The VDV values were all above 1.3 for all axes of motion, generally indicating high levels of shocks and a risk of health problems (see Figure 8.12). The mean seat effective amplitude transmissibility factor (SEAT[13]) indicated a general ineffectiveness of any of the seat suspension systems. The authors concluded that "…these data indicate that locomotive rides are characterized by relatively high shock content (acceleration peaks) of the vibration signal in all directions. Locomotive vertical and lateral vibrations are similar, which appears to be characteristic for rail vehicles compared with many road/off-road vehicles. Tested locomotive cab seats currently in use (new or old) appear inadequate to reduce potentially harmful vibration and shocks transmitted to the seated operator, and older seats particularly lack basic ergonomic features regarding adjustability and postural support."

Kress et al. (2009) examined locomotive seats and found that they were deficient in lumbar support, vibrational characteristics, vertical and fore/aft adjustment control design, and biomechanical soundness. Vibration measurements showed high crest factors (indications of shock transients). They measured low frequency rolling around the x axis (see Figure 8.9) that caused body swaying in occupants which the occupants constantly attempted to counteract. This could cause fatigue in the muscles that support the spinal column and increase disk pressure and injury risk.

The Multer et al. (1998) report cites the information in ISO 2631 discussed above and recommends the use of active systems to control vibration and dampers on the seat post to reduce vibration exposure to the train crew. Seats conduct vibration to the occupant, so seat design is important ergonomically. The report has a separate section on Seating which will be discussed below in the Cab Ergonomics section.

WALKING ON BALLAST

Reinach and Gertler (2001) conducted a broad examination of safety issues in the yard and terminal environment of railroads. They reported that one-third of railroad worker injuries occurred in railroad yards. They found that "Sprains and strains accounted for more than half of the LWD [lost work days] injuries in railroad yards; the trunk/torso was the most affected body part; slips, trips and falls were the most common triggering event; and the acts of walking, running, or stepping over were the leading physical acts associated with LWD yard injuries." (p. 2). A significant contributor to these injuries was the use of large ballast in an environment where employees need to walk to perform their duties, such as operating switches. In the period for which FRA injury data was examined (1997–1998), ballast was involved in 304 injuries, or 6.6%. Moreover, ballast-related injuries resulted in the most severe injuries with a median of 30 lost workdays. The use of improper ballast was among the factors cited in their report that contributed to yard injuries. They suggested the use of smaller "walking" stone on switch leads instead of larger ballast rock.

Wade et al. (2010) examined the same issue in the FRA database and found that "…between 1998 and 2006, walking contributed to 13.9% to 16.5% of all railroad worker injuries and accounted for 16.7% to 20.3% of the days absent from work during the same period (FRA, 1999–2008). In addition, slip, trip, and fall (attributed to irregular surface contact) injuries accounted for between 3.1% and 5.1% and represented between 3.5% and 5.9% of the days absent from work during the period (FRA, 1999–2008). The FRA also found that in 2008, 35% of all injuries reported were associated with torso and lower extremity sprains and strains (FRA, 1999–2008).

Thus, the question arises about the potential link between walking surface characteristics and injury." (p. 560).

Wade et al. studied 20 adult men who walked on three distinct pathways (no ballast, walking ballast, and mainline ballast). Walking ballast rock had an average size of 0.75 to 1 inch, and mainline ballast had rock with an average size of 1.25–1.5 inch. They recorded full-body motion, ground reaction forces, and electromyographic (EMG) signals.

They found that joint moments were smaller for mainline ballast and walking ballast when compared with no ballast. EMG activity was found to be significantly greater on both types of ballast. Moreover, temporal gait parameters (cadence, speed, stride length) were significantly different for mainline ballast than for either walking ballast or no ballast. They concluded that walking on ballast increases muscle activity to control the lower extremity joints, and that "…these findings provide preliminary support for the concern of "improper ballast" as a factor that contributed to injuries in railroad yards and suggest that the "distribution of smaller 'walking' stone on switch leads instead of the larger ballast rock" might help reduce railroad yard injuries." (p. 569). A less technical version of this research can be found in Jones and Andres (2011).

It should be noted that the research by Wade et al. was strongly opposed by the AAR and the railroads. For details see Chapter 2.

CAB ERGONOMICS

Most of the issues concerned with working conditions within the locomotive cab environment have been addressed in sections of this chapter above such as noise, temperature, vibration and sanitation. This section focuses on the layout and general design of the locomotive cab. These issues were critically and comprehensively address in an FRA report by Multer et al. (1998). There has not been a subsequent report that addresses these issues. A brief summary of the report is provided below.

General Design

The layout of the locomotive cab should be designed using population sizes with the male 95th percentile dimensions to set clearances and the female 50th percentile to reach envelopes. The height of the cab ceiling should be at least 76 inches. Sufficient floor space should be allowed for each occupant with 65 sq. ft. as a minimum. The toilet, water cooler, storage, refrigerator and other "amenity" items should be located out of the main area of the cab and not counted as crew space. Regular viewing tasks should be within a 30° cone around the normal line of sight (10–15° below the horizontal plane) to minimize fatigue due to poor neck and head posture. A viewing angle between 5° above and 30° below the horizontal plane should be used to establish the height of the seat relative to the windows and the visual displays in the cab.

Location of Controls and Displays

Primary displays and controls should be viewed by the engineer without changing eye or head position from the normal line of sight. Secondary displays and control should be placed so that head movements are not necessary, but eye movements are. Control movements should be consistent with the display or system response. Controls should be spaced and sized to allow operation without accidental activation of adjacent controls.

Access

Doors should be located in consideration of post-accident evacuation and should open outward. A roof hatch or detachable side window should be considered as an alternate evacuation exit.

Seating

Locomotive seats should be designed with a minimum cushion 3 inches thick with arm rests 4 inches wide and 13 inches long. Seats should be adjustable in height within the range from 16 to 19 inches and be adjustable forward and back by 4 inches within the 50th percentile position. Seating comfort considerations include leg room, knee room, availability of footrests, clearance from sidewall, vibration levels, and ease of entry/exit from the seat.

Other General Design Issues

The report also has guidelines for workstation design, electromechanical displays, auditory devices, alarms, computer generated displays, and automation.

NOTES

1 dBA refers to a sound level measurement that is weighted to "...an approximation of equal loudness perception characteristics of human hearing for pure tones relative to a reference of 40 dB SPL at 1 kHz..." (Berger et al., 2000, p. 54). SPL is the abbreviation for sound pressure level.

2 A requirement of the Federal Railroad Safety Act of 1970 was an annual report to Congress. Apparently, several of these reports have been lost as they cannot be found on the FRA website.

3 It is not clear what Lav represents. The abbreviation Lav is not defined in current texts and is not defined in the 1996 report. It would seem to be most likely the same as L_A in current terminology, which is the A-weighted sound level (Berger et al., 2000, p. xv). A definition of TWA can be found in Berger et al. (2000, p. 74).

4 In this system, a plastic seat is placed over a 5-gallon bucket with a plastic bag. After use, the bag is tied and, upon arrival at a final terminal, the employee places the bag at a holding facility and it is disposed of later. The urinal consists of two parts. The upper portion is used for the washing of hands while the lower portion is used for urination. Both fluids are drained directly to the ground." (FRA, 1996, pp. 8–5)

5 A z-score is a normalized score. For instance, if the mean of a distribution of scores (temperatures) is 51.5 and the SD is 20, then the z-score for a temperature of 100 is $(100 - 51.5)/20 = 2.425$.

6 z-score probabilities can be looked up in tables of the cumulative normal distribution. Many spreadsheet programs can calculate probabilities for the cumulative normal distribution. A z-score of 2.425 has a probability, $0.9922397 < p_z < 0.9924506$, based on a table of the cumulative normal distribution.

7 A meta-analysis is a statistical analysis that combines the results of multiple research studies.

8 This is the change in sound level, in decibels, which would require halving or doubling of the allowable exposure time to maintain the same noise dose, in this case, TWA 90 dB(A).

9 This is the limit for employee noise exposure. It is the exposure level at which hearing protection, engineering or administrative controls must be used to limit noise exposure.

10	For example, the European Union's *Guide to Good Practice on Whole-Body Vibration* (Anonymous, 2008, p. 14) shows vibration magnitudes for more than 20 common tools (e.g., bulldozers, dump trucks), but not for railroads. A literature review for whole-body vibration in construction and other vehicles (Kittusamy and Buchholz, 2004) only references one article for subway train drivers. Griffin (1990, Appendix 5) cites only six of 135 field studies of health issues caused by vibration that involved railroad employees.

11	Maximum transient vibration value. This is a calculation that evaluates shock-type vibrations (Marquis, 2009, p. 12).

12	The measurements and data analysis reported in this paper were supported in part by the Union Pacific Railroad and the CSX Railroad.

13	See Griffin (1990) pp. 405–408 for the technical definition of SEAT. Griffin (1990, p. 846) defines SEAT as "A non-dimensional measure of the efficiency of a seat in isolating the body from vibration or shock."

REFERENCES

Anonymous. (2008). *Guide to good practice on whole-body vibration.* EU Good Practice Guide WBV.

Berger, E. H., Royster, L. H., Royster, J. D., Driscoll, D. P., & Layne, M. (2000). *The noise manual* (5th edn). Fairfax, VA: American Industrial Hygiene Association.

British Standards Institution. (1987). *Measurement and evaluation of human exposure to whole-body mechanical vibration and repeated shock.* British Standard, BS 6841.

Brooke, S., & Ellis, H. (1992). Cold. In D. M. Jones & A. P. Smith (Eds.), *Handbook of human performance: The physical environment* (Vol. 1, pp. 105–130). London: Academic Press.

Committee for Study of the Federal Employers' Liability Act. (1994). *Compensating injured railroad workers under the Federal Employers' Liability Act* (Special Report 241). Washington, DC: National Academy Press.

Dunn-Rankin, P. (1983). *Scaling methods.* Hillsdale, NJ: Lawrence Erlbaum Associates.

Federal Railroad Administration, Office of Research and Development. (1992). *Improving transportation through railroad research* (Report No. DOT/FRA/ORD-92/14). Washington, DC: U.S. Department of Transportation.

Federal Railroad Administration, Office of Safety Assurance and Compliance. (1996). *Locomotive crashworthiness and cab working conditions* (Report to Congress). Washington, DC: U.S. Department of Transportation.

Federal Railroad Administration. (1976). *1976 annual report by the President to the Congress on the administration of the Federal Railroad Safety Act of 1970.* Washington, DC: U.S. Department of Transportation. https://www.fra.dot.gov/eLib/details/L19670#p3_z5_gD_kFederal%20Railroad%20Safety%20Act.

Federal Railroad Administration. (1977). *1977 annual report by the President to the Congress on the administration of the Federal Railroad Safety Act of 1970.* Washington, DC: U.S. Department of Transportation. https://www.fra.dot.gov/eLib/details/L19671#p2_z5_gD_kFederal%20Railroad%20Safety%20Act

Federal Railroad Administration. (1981). *1981 annual report by the President to the Congress on the administration of the Federal Railroad Safety Act of 1970.* Washington, DC: U.S. Department of Transportation. https://www.fra.dot.gov/eLib/details/L19668#p2_z5_gD_kfederal%20railroad%20safety%20act

Federal Railroad Administration. (1982). *1982 annual report by the President to the Congress on the administration of the Federal Railroad Safety Act of 1970.* Washington, DC: U.S. Department of Transportation. https://www.fra.dot.gov/eLib/details/L19665#p2_z5_gD_kfederal%20railroad%20safety%20act

Federal Railroad Administration. (1983). *1983 annual report by the President to the Congress on the administration of the Federal Railroad Safety Act of 1970*. Washington, DC: U.S. Department of Transportation. https://www.fra.dot.gov/eLib/details/L19666#p2_z5_gD_kfederal%20railroad%20safety%20act

Federal Railroad Administration. (1984). *1984 annual report by the President to the Congress on the administration of the Federal Railroad Safety Act of 1970*. Washington, DC: U.S. Department of Transportation. https://www.fra.dot.gov/eLib/details/L19667#p1_z5_gD_kfederal%20railroad%20safety%20act

Federal Railroad Administration. (2002). 49 CFR Part 229. Locomotive cab sanitation standards. *Federal Register, 67,* 16032–16052.

Federal Railroad Administration. (2006). Occupational noise exposure for railroad operating employees. 49 CFR Part 227 and 229. *Federal Register, 71,* 63065–63138.

Federal Railroad Administration. (2012). 49 CFR Part 229 and 238. Locomotive safety standards. *Federal Register, 77,* 21312–21357.

Gamst, F. C. (1975). Human factors analysis of the diesel-electric locomotive cab. *Human Factors, 17,* 149–156.

Gertler, J., & DiFiore, A. (2009). *Work schedules and sleep patterns of railroad train and engine service workers* (Report No. DOT/FRA/ORD-09/22). Washington, DC: U.S. Department of Transportation.

Griffin, M. J. (1990). *Handbook of human vibration*. New York: Academic Press.

Griffin, M. J. (1992). Vibration. In D. M. Jones & A. P. Smith (Eds.), *Handbook of human performance: The physical environment* (Vol. 1, pp. 55–78). London: Academic Press.

Griffin, M. J. (1997). Vibration and motion. In G. Salvendy (Ed.), *Handbook of human factors and ergonomics* (2nd edn, pp. 828–857). New York: Wiley.

Guilford, J. P. (1954). *Psychometric methods* (2nd edn). New York: McGraw-Hill.

Heinrich H.W. (1959). *Industrial accident prevention: a scientific approach* (4th edn). New York: McGraw-Hill.

Hobbs, J.R., Walter, R.A., Hard, T., & Devoe, D. (1977). *Train generated air contaminants in the train crew's working environment* (Report No. FRA/ORD-77/08). Washington, DC: U.S. Department of Transportation.

Hunter, J. E., & Schmidt, F. L. (1990). *Methods of meta-analysis: Correcting error and bias in research findings*. Newbury Park, CA: Sage Publications.

Hygge, S. (1992). Heat and performance. In D. M. Jones & A. P. Smith (Eds.), *Handbook of human performance: The physical environment* (Vol. 1, pp. 79–104). London, UK: Academic Press.

International Organization for Standardization. (1996). *ISO 9996 Mechanical vibration and shock-disturbance to human activity and performance- Classification*. Geneva, Switzerland: Author.

International Organization for Standardization. (1997). *ISO 2631-mechanical vibration and shock: Evaluation of human exposure to whole-body vibration*. Geneva, Switzerland: Author.

Johanning, E., Fischer, S., Christ, E., Göres, B., & Landsbergis, P. (2002). Whole-body vibration exposure study in U.S. railroad locomotives: An ergonomic risk assessment. *American Industrial Hygene Associaton Journal, 63*(4), 439–446. doi:10.1080/15428110208984732

Johanning, E., Landsbergis, P., Fischer, S., Christ, E., Gores, B., & Luhrman, R. (2006). Whole-body vibration and ergonomic study of US railroad locomotives. *Journal of Sound and Vibration, 298,* 594–600.

Joint Committee on the *Standards for Educational and Psychological Testing* of the American Educational Research Association, the American Psychological Association, and the National Council on Measurement in Education. (2014). *Standards for educational and psychological testing*. Washington, DC: American Educational Research Association.

Jones, M. E., & Andres, R. (2011). *Joint loading and muscle activity in the lower body while walking on ballast* (RR 11–04). Washington, DC: U.S. Department of Transportation.

Kahneman, D., Slovic, P., & Tversky, A. (Eds.). (1982). *Judgment under uncertainty: heuristics and biases*. New York: Cambridge University Press.

Kilmer, R. D. (1980). *Assessment of locomotive crew in-cab occupational noise exposure* (Report No. DOT/FRA/ORD-80/91). Washington, DC: U.S. Department of Transportation.

Kittusamy, N. K., & Buchholz, B. (2004). Whole-body vibration and postural stress among operators of construction equipment: A literature review. *Journal of Safety Research, 35,* 255–261.

Krantz, D. H., Luce, R. D., Suppes, P., & Tversky, A. (1971). *Foundations of measurement: Additive and polynomial representations* (Vol. 1). New York: Academic Press.

Kress, T. A., Kroman, A. M., & Mullinix, L. (2009). *Ergonomic assessment of locomotive seats.* Unpublished manuscript.

Kryter, K. D. (1985). *The effects of noise on man* (2nd edn). Orlando, FL: Academic Press.

Licklider, J. C. R. (1951). Basic correlates of the auditory stimulus. In S. S. Stevens (Ed.), *Handbook of experimental psychology* (pp. 985–1039). New York: Wiley.

Lindsay, R. B. (1981). Acoustics. In H. L. Anderson (Ed.), *Physics vade mecum.* New York: American Institute of Physics.

Marquis, B. (2009). *Locomotive ride safety.* Unpublished PowerPoint presentation.

Multer, J., Rudich, R., & Yearwood, K. (1998). *Human factors guidelines for locomotive cabs* (Report No. DOT/FRA/ORD-98/03). Washington, DC: U.S. Department of Transportation.

Peay, J. M., & Sanders, M. S. (1978). *Health and safety implications of diesel locomotive emissions* (Report No. DOT/FRA/ORD-78/18). Washington, DC: U.S. Department of Transportation.

Pilcher, J. J., Nadler, E., & Busch, C. (2002). Effects of hot and cold temperature exposure on performance: a meta-analytic review. *Ergonomics, 45,* 682–698.

Ramsey, J. D., Burford, C. L., Beshir, M. Y., & Jensen, R. C. (1983). Effects of workplace thermal conditions on safe behavior. *Journal of Safety Reseach, 14,* 105–114.

Raslear, T. G. (1998/2020a). *Cold weather temperature in locomotive cabs.* Unpublished manuscript.

Raslear, T. G. (1998/2020b). *On the relationship between temperature and accidents.* Unpublished manuscript.

Raslear, T. G. (2003). Effects of hot and cold temperature exposure on performance: a meta-analytic review. Unpublished PowerPoint presentation.

Reason, J. (1990). *Human error.* Cambridge, UK: Cambridge University Press.

Reason, J. (1997). *Managing the risks of organizational accidents.* Aldershot: Ashgate.

Reinach, S., & Gertler, J. (2001). *An examination of railroad yard worker safety* (Report No. DOT/FRA/ORD-01–20). Washington, DC: U. S. Department of Transportation.

Remington, P. J., & Rudd, M. J. (1976). *An assessment of railroad locomotive noise* (Report No. DOT/FRA/ORD-76/142). Washington, DC: U.S. Department of Transportation.

Smith, S. D., Smith, J. A. & Newman, R. J. (2006). *Vibration transmissibility characteristics of occupied suspension seats* (Report No. AFRL-HE-WP-TR-2006-0 133). Wright-Patterson AFB, OH: Air Force Material Command.

Strout, P. E., & Lane, S. P. (2012). Reliability. In H. Cooper, P. M. Camic, D. L. Long, A. T. Panter, D Rindskopf, & K. J. Sher (Eds.), *APA handbook of research methods in psychology: Foundations, planning, measures, and psychometrics* (Vol 1., pp. 643–660). Washington, DC: American Psychological Association.

Stusnick, E., Montroll, M., Plotkin, K., & Kohli, V. (1982). *Railroad noise control. Handbook for the measurement, analysis and abatement of railroad noise* (Report No. DOT/FRA/ORD-82/02). Washington, DC: U.S. Department of Transportation.

Thurlow, W. R. (1971). Audition. In J. W. Kling and L. A. Riggs (Eds.), *Woodworth & Schlosberg's experimental psychology* (3rd edn, pp. 223–271). New York: Holt, Rinehart and Winston.

Tukey, J. W. (1977). *Exploratory data analysis*. Reading, MA: Addison-Wesley.

Wade, C., Redfern, M. S., Andres, R. O., & Breloff, S. P. (2010). Joint kinetics and muscle activity while walking on ballast. *Human Factors: The Journal of the Human Factors and Ergonomics Society, 52,* 560–573.

9 Safety Culture Programs

BACKGROUND

There was an early realization by Mike Coplen that the safety culture of the railroad industry — the carriers, labor, and the FRA — was a major problem in the causation of accidents. Mike's prior employment in the industry as a railroad engineer provided him with this insight which he shared with me and several colleagues at the Volpe Center. Eventually, safety culture programs with three different foci were devised and promulgated in the industry. These programs have largely been incorporated into the industry and persist as important components of safety at many railroads.

When the Human Factors Research Division first started to talk about safety culture in public and private meetings with the carriers, labor and other components of the FRA, the question was often asked "what is safety culture"? Many thought safety culture was just another catchphrase with no meaning and no real applicability to the practical problems of the railroad industry. Accordingly, safety culture was sometimes dismissed as a frivolous concept and was often ridiculed. However, seminal work by Reason (1997), Leveson, (1995), Moray, (1994) and others provided a firm basis for arguing that safety culture was real and had practical value for the railroad industry.

What Is Safety Culture?

Ranney (2003) provided an overview of safety culture concepts at the Transportation Research Board Human Factors in Transportation Workshop. She noted that every organization has a culture which involves learned behaviors, shared values and assumptions that allow the organization to cope with internal and external issues. There are subcultures that address safety, customer service and quality. According to Reason (1997), safety culture is "Shared values (what is important) and beliefs (how things work) that interact with an organization's structures and control systems to produce behavioral norms (the way we do things around here)." Changing safety culture in a positive direction involves action on several levels. First, there are unsafe acts on the part of frontline employees. Often such acts result in the discipline of the employee, and no further action is taken by the organization. However, there are often workplace factors which are the conditions and tools that affect the workers performing the task. If these are not addressed, unsafe acts will still occur. Workplace factors need to be addressed at the organizational level. Organization factors represent the safety and business climate of the workplace, such as values and priorities, resource allocations, and organizational structures and processes. In other words, a systems approach[1] is needed to change safety culture. A systems approach recognizes that individuals act in a context of workplace factors that are promulgated by an organization's safety and business climate, which, in turn, is influenced by the social, legal, and policy context outside the railroad industry. Finally, Ranney suggests several interventions to change safety culture. These include safety rules

DOI: 10.1201/9781003500018-13

revision, behavior-based safety, and close call reporting systems. FRA research projects on safety rules revision, behavior-based safety and close call reporting systems are described and discussed in sections below on Safety Rules Revision, Behavior Based Safety, and Close Call Reporting Systems.

The characteristics common to all three of these programs are that they increase the organization's value of safety, promote upward safety communication, increase team work, increase the tendency to approach others with safety concerns, and increase injury reports. From a systems point of view, these programs promote interaction between the layers of the sociotechnical system and allow safety problems to be recognized, addressed and resolved.

SAFETY RULES REVISION

Safety rules at railroads are often said to be written in blood. This is because investigations of serious accidents and fatalities frequently result in a determination that operator error was the proximal cause of the accident, and the obvious solution thus becomes a rule that mandates that operators perform their duties in a prescribed fashion designed to prevent the occurrence of that particular error. There are multiple problems with this approach to improving safety and avoiding similar accidents.

First, this approach usually ignores the work conditions that were responsible for the operator action(s) that resulted in the accident. If an employee ran through a switch and derailed, mandating a speed restriction and crew acknowledgement of the switch position may not improve safety if, for example, the employees were fatigued and experiencing microsleeps. Organizational policies concerning work conditions are often required to fix a situation that "forces" employee actions that result in an accident. For example, changing company policies about work schedules might allow fatigued employees to refuse a job and avoid operating when fatigued.

Second, blaming the person at the sharp end of an operation leads to a lack of communication between management and operators. There are often incidents in which there is no tangible damage or personal injury. If such incidents are reported, they result in employee discipline and, consequently, they are not reported so that potentially dangerous conditions are not seen by management and are not addressed before a serious incident occurs.

Third, because there are many safety-related incidents in any operational organization, promulgating new rules based on serious incidents results in rule books that are complex, have conflicting rules, and are difficult to remember and interpret.

There are many reasons why safety rules proliferated in the railroad industry (FRA, 2003). As noted above, it may be because management wishes to reduce accidents. It may be because the railroads wish to limit liability. Under the Federal Employers Liability Act (FELA[2]), plaintiffs must show fault by the railroad to receive compensation. If an employee violates a rule, the railroad cannot be considered at fault. Finally, in the 1980s and 1990s there were many railroad mergers that resulted in a large number of overlapping and conflicting safety rules. Regardless of the cause, many safety proponents in the railroad industry perceived the need for a process to reduce the number of safety rules and to revise the rules for clarity and ease of application.

Operating and safety rules have long been recognized as an issue by FRA. Devoe and Story (1973) first addressed the problem of writing or revising rules to avoid confusion and ambiguity due to the wording of rules. Mike Coplen (Coplen, 1999), an FRA research manager, suggested that safety rules revisions that produced fewer, more carefully crafted rules, would reduce confusion about the scope and application of rules. Moreover, if the process involved both labor and management, the process would build trust between labor and management through broad workforce participation in the writing of rules and improve compliance because of increased workforce ownership of the new rules. Thus, it was suggested that rules consolidation would lead to fewer injuries and improvements in safety culture. As an added benefit to the railroads, rules revisions might also reduce liability by clarifying which rules are applicable 100% of the time. The cost of injuries in the industry at the time (FRA, 2003) was substantial, and a reduction in both injuries and liability would be a substantial benefit to the railroads.

Ranney and Nelson (2004, 2007) documented the effects of participatory safety rules revisions in three railroad companies (CSX Transportation, Kansas City Southern, Canadian National/Illinois Central) and a barge line (American Commercial Barge Lines). They assessed the effectiveness of these programs on safety culture, incident rates and liability costs. In addition to statistical analyses of incident rates, they conducted interviews with railroad safety executives and railroad workers/union representatives (Figure 9.1).

The far left box in the figure shows the conditions that lead to a need for rules revision. These include a high volume of rules stemming from injury avoidance, mergers and liability concerns as noted above. The activities in the safety rules revision process are shown in the next box and include changing the roles and responsibilities of management and workers regarding the drafting of rules. Rules revision in this model requires that management, employees and unions work together on all aspects of the process. "A Core Team is created to oversee the work of craft-specific Satellite Teams, draft rules and non-rules materials (e.g., recommended work practices), disseminate information, incorporate feedback, and promote the cultural shift." (Ranney and Nelson, 2007, p. 15). Craft-specific teams are

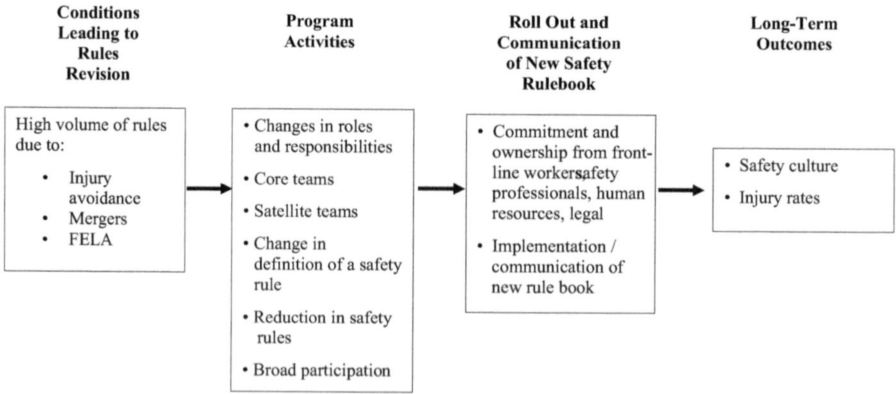

FIGURE 9.1 Rules revision program theory.

created to draft craft-rules, provide feedback and promote the cultural shift. The idea is to generate broad participation in the process while reducing the number of rules. The third box concerns compliance with the new rules. Ownership (especially by employees), cooperation and communication between parts of the organization are important for increased compliance. The long-term goals and outcome measures of rules revision are shown in the fourth box. These include a reduction in incident rates, a reduction in liability costs, and improvements in safety culture. According to Ranney and Nelson, 2007, p. 15): "The process changes the 'values, attitude, competencies and patterns of behavior' by shifting the primary responsibility for drafting rules from management to workers and by revising or eliminating rules that are not meaningful to the workforce. When rules are conflicting and voluminous, one can easily defend noncompliance to oneself and to one's peers. When only a few rules exist that one's peers have drafted personally, common sense and peer pressure encourage compliance. Rules revision, in short, is a process of 'streamlining rules into the culture.'"

Ranney and Nelson (2007, p. 28) reported the following outcomes:

Changes in values, attitudes, and patterns of behavior related to safety rules
Changes in the usability and number of safety rules
Changes in union-management relations
Perceived changes in compliance
Changes in incident rates
Changes in injury-related liability and costs
Spillover effects in non-safety areas

Ranney and Nelson (2007) note several examples of positive changes in safety culture. For instance, there was a change in the roles of management and labor regarding what rules were credible. Also, each craft was now responsible for its own rules. A new definition of a rule was formulated and what could be included in drafting their own rulebook, and there was increased concurrence about every rule in the rule book. These changes all go to changes in values, attitudes, and patterns of behavior that are the key components of safety culture.

The primary goal of participatory rules revision is the reduction of the number of rules. For Kansas City Southern there were 742 rules before the revision. After the process, there were 17 core rules and 660 craft-specific rules. Similar results were reported for CSX Transportation and the barge line.

Interview data indicated that the rules revision process had been associated with improvements in rules compliance. Safety executives reported that they observed fewer at-risk behaviors among employees. However, statistical data were not available to evaluate this change.

A key indicator of success for rules revision as a safety intervention is the extent to which it resulted in a reduction in incidents. Data only existed for two of the railroads and the barge line. Unfortunately, the data did not reveal a convincing picture of improvement in incidents as can be seen in Figure 9.2 which shows the incident rate for Kansas City Southern and the industry from 1995 to 2002. For this carrier

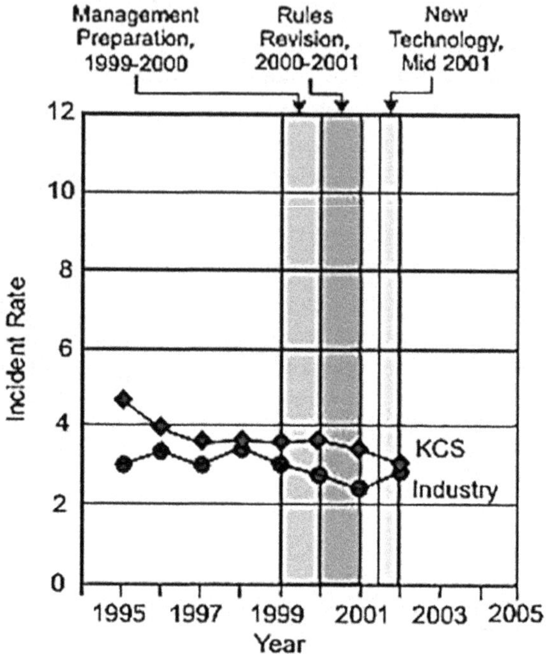

FIGURE 9.2   Comparison of KCS and industry incident rates.

the rate of decrease in incident rate relative to the industry was statistically reliable. However, a similar trend was not seen in the other carriers. Ranney and Nelson provide several possible reasons for this, but the data do not, generally, support rules revision as a rigorous safety intervention.

A decrease in the number of claims and the claims payouts was noted by two of the railroads. However, Ranney and Nelson were not able to confirm this with documentation. The other carriers either did not respond to this issue or had no data that could be shared.

Overall, the studies cited here do not provide conclusive evidence that participatory safety rules revision programs are effective mechanisms for positive changes in safety culture or a reduction in incidents. Part of the problem with this research is that the researchers depended on finding carriers who were implementing a rules revision program that they could document with interviews and statistical data. Unfortunately, that did not provide the researchers with the level of control over the rules revision process that might have ensured a more conclusive outcome. In other words, they did not control the process, they merely documented it. Also, they did not have the ability to formulate and implement a rigorous data acquisition plan because of the *ad hoc* nature of the research. Finally, there were too few interviews and sparse data collected to allow more robust qualitative and quantitative analyses. Other safety culture projects discussed below were extensively planned and controlled by the researchers to ensure conclusive outcomes, whether positive or negative.

BEHAVIOR BASED SAFETY: CLEAR SIGNAL FOR ACTION

The Clear Signal for Action (CSA) project was initiated in 2001 using the ideas of the broad area of Behavior Based Safety (Coplen, 2014a, b). Behavior Based Safety (BBS) is the application, in part, of the principles of The Experimental Analysis of Behavior as formulated through experimentation by B. F. Skinner and his associates (Skinner, 1938; Keller and Schoenfeld, 1950; Reynolds, 1975). Operant behaviors (i.e., behaviors that are not reflexes or involuntary) are elicited by their antecedents and are influenced (increased or decrease in occurrence) by their consequences. This is often called the ABC model (Krause, 1995; Raslear, 2011):

Antecedents: Behavior → Consequence

Antecedents elicit behavior, they do not cause the behavior. The operant behavior is not reflexive, it is voluntary. Antecedents are said to "set the stage" for the behavior. In the context of railroad safety, for example, the behaviors are various human "errors", the consequences are various accidents or incidents, and the antecedents maybe the physiological state of the railway employee (e.g., fatigue), the availability of personal protective equipment (PPE), or the state of repair of the infrastructure or rolling stock. To restate the general ABC model for the railroad industry:

Fatigue, No PPE, Poor Maintenance: Errors → Accident, Incident

More specifically, using the example of fatigue-caused accidents as noted in Chapter 7, the ABC model might look like this:

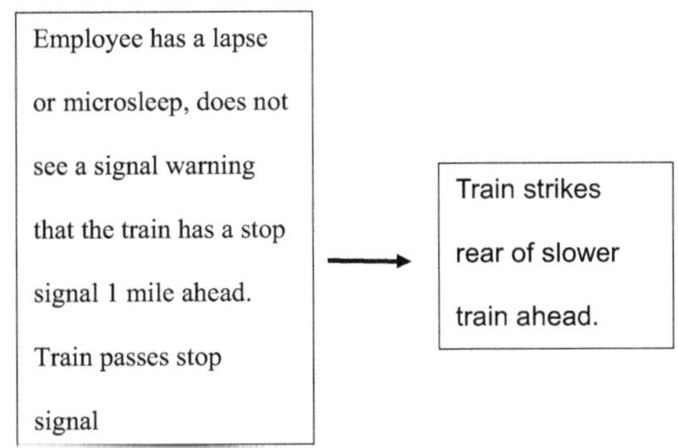

All errors (operant behaviors) have antecedents and consequences. In the railroad industry the consequences of interest are accidents and incidents. The antecedent of interest may be a physiological state such as fatigue. The behaviors are various human "errors". The errors are revealed in post-accident investigations such as a signal passed at danger (perhaps elicited by a lapse).

The contractor responsible for implementing CSA was Behavioral Science Technologies (BST). In addition to Applied Behavior Analysis (peer-to-peer observations and feedback), as described in a simplified form above, BST incorporated Quality Management (continuous process improvement) and Organization Development (safety leadership development) in its processes for improving safety in organizations (Coplen, 2014a, b; Krause, 1995, 1996, 2002, 2005; Krause and Bell, 2015).

Peer-to-peer observations and feedback constitute a fundamental aspect of the BST approach (Krause, 1995, 1996). The emphasis is on the ABC model in which peer-to-peer observations focus on targeted at-risk behaviors. The at-risk behaviors are stated in objective, observable terms such as "failure to wear respiratory protective equipment". The antecedents of this behavior are then developed from discussions and interviews with workers. In the case of respiratory PPE, this might include the availability of the PPE, time of day, peer pressure, lack of training and poor modeling of the behavior by supervisors who occasionally do not wear respiratory PPE, etc. An important antecedent is the attitudes of the workforce regarding the at-risk behavior. Attitudes consist of beliefs, subjective norms and perceived control. Then the consequences of the at-risk behavior are described. In a behavioral analysis the most effective consequences (according to Krause) are said to be "soon[3]", "certain[4]" and "positive[5]". In the respiratory PPE example, consequences include "saves time", "injury or death", "reprimand", "comfort" and "convenience". Consequences which are "late, uncertain and negative" are the weakest and include "injury or death" and "reprimand". Thus, the ABC analysis shows that the consequences are biased toward reinforcing the at-risk behavior of not wearing respiratory PPE because the consequences are "soon, certain and positive". Peer-to-peer observations then commence in which safe behaviors and at-risk behaviors are recorded. Positive feedback is provided for the safe behaviors observed, and behaviors that need improvement are discussed. The consequences of the behavior provide feedback in a closed loop, which influence the antecedents, including skills (how to properly wear PPE), facility conditions (PPE is easily available where needed), and attitudes about the use of PPE.

The working interface, according to Krause (2005) "… is the configuration of equipment, facilities, systems, and behaviors that define the interaction of the worker with the technology." (p. 10). The working interface is an important focus of human factors as noted in Chapter 1. Workers are exposed to safety hazards in the working interface. Improved safety results from controlling worker exposure to those hazards. This is not a matter of unsafe worker behavior or of malfunctioning equipment. Rather, the interaction of the worker with the technology is crucial to understanding why accidents occur and how to prevent them. Variables that influence the working interface include the quality of the design of the system and its elements, training, organizational safety culture (beliefs, subjective norms and values) and leadership. Understanding the working interface is a key element of BBS and is achieved by peer-to-peer observations and feedback as noted above. Quality Management (continuous process improvement) tracks the process by which critical behaviors are identified, their root causes (antecedents) are identified, and an action plan to reduce or eliminate the critical behaviors is implemented. Safety performance (such as occurrences of unsafe behaviors) is tracked, and if improvement does not occur, the process is repeated until a satisfactory safety outcome is achieved (see Krause, 1997, Figures 3–8).

It is obvious that BBS can only work if the employees are engaged in the processes of peer-to-peer observations and feedback. However, Krause (2005) also emphasizes leadership in enabling safety systems and sustaining those systems in the organization. Safety systems are mechanisms that enable safety in the working interface. They may include hazard recognition and mitigation, training, procedures and policies. These systems must be sustained by leadership if safety is to be maintained

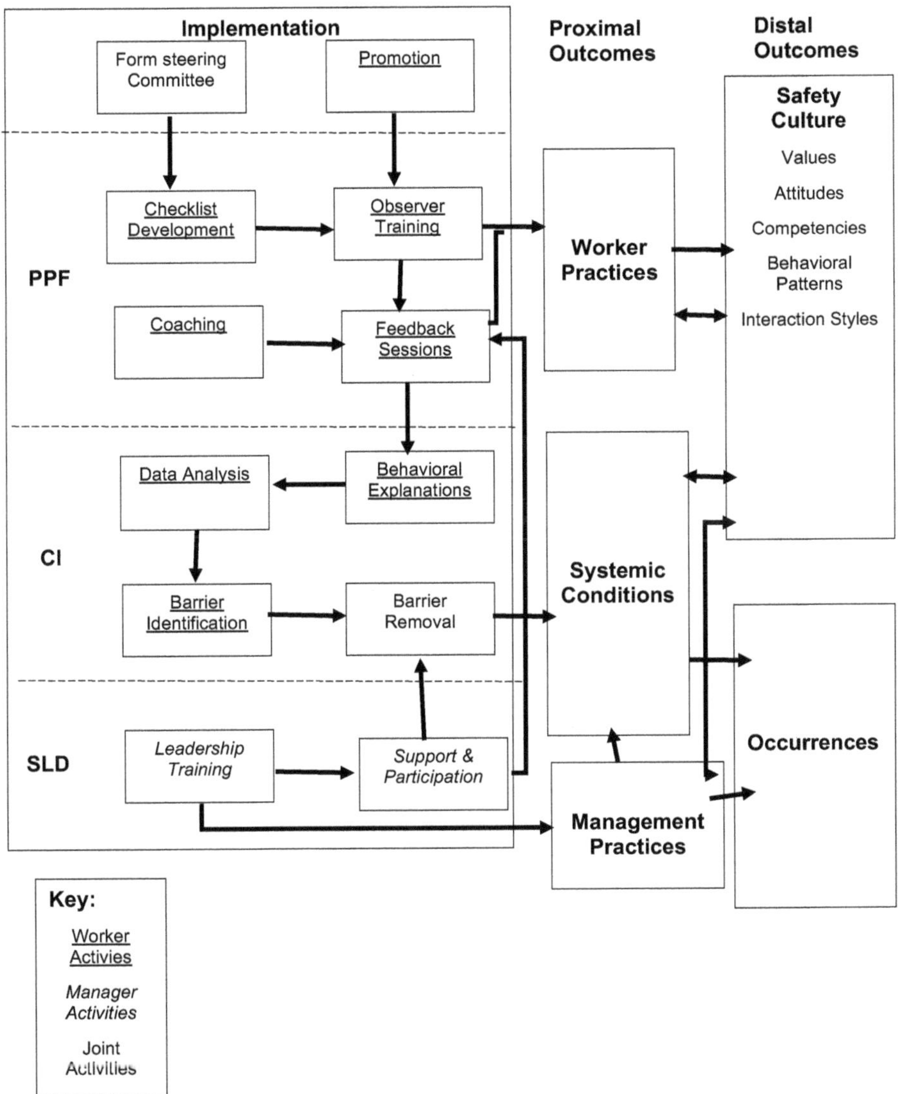

FIGURE 9.3 Theory of action for CSA, showing activities and theoretical outcomes (Zuschlag et al. 2012).

and improved. Personnel selection and development, performance management, and employee engagement are processes that organizational leaders may use to sustain safety systems. All of this (safety enabling systems, the sustaining of safety systems, and the working interface) constitutes the driving values of the organization — "how we do things here". Leadership can be said to create the organization's safety culture. Consequently, Organization Development (safety leadership development) is an important aspect of BST's approach to BBS.

Figure 9.3 shows the Theory of Action for CSA as described by Zuschlag et al. (2021). As noted above, the project involves three elements:

Peer-to-peer Feedback (PPF). PPF uses behavioral analysis methods to identify and address at-risk behaviors before they cause injuries by using peer-to-peer observations of, and two-way feedback about, work behavior, conditions, and organizational factors.

Continuous Improvement (CI). In a safety context, CI establishes a means of gathering data to identify systemic causes of observed at-risk behaviors and conditions, and then implements corrective actions to address the causes. Potential systemic causes include organizational policy, training, tool design, environmental conditions, procedures, and cultural aspects. Data gathering continues after a corrective action is taken, allowing time to evaluate its effectiveness.

Safety Leadership Development (SLD). SLD provides management training to promote proactive safety practices, such as PPF and CI.

The Outcome

Starting in 2001, two Class I railroads participated in the CSA project: Amtrak and Union Pacific. Union Pacific had projects at two sites. The projects were given names by the employees at each site. The Amtrak site was the Chicago terminal and was called EAGLES (Employee Alliance for Great Levels of Excellence in Safety). The EAGLES site involved baggage, Red Caps, ticket and gate agents, and customer service (Federal Railroad Administration, 2007a). The Union Pacific site at the San Antonio Service Unit (SASU) of Union Pacific Railroad (UP) was called Changing At-risk Behavior (CAB). CAB (Federal Railroad Administration, 2008a, b, 2009a) was initiated in 2005 and involved road and yard crews (engineers and conductors). A second site, called STEEL (Safety Through Employees Exercising Leadership) was established at the Livonia Service Unit (LVSU) and applied the same method to managers and switching-yard crews in Spring 2006 (Federal Railroad Administration, 2009b).

Ranney et al. (2013) provide a summary of all of the FRA's safety culture programs over a 14-year period, including CSA. The improvements in safety are impressive as can be seen in Table 9.1.

TABLE 9.1

Safety Improvements Under the CSA Project

Railroad	Site	Employee Group	Outcome
AMTRAK	Chicago Terminal	Baggage, Customer Service, etc.	76% decrease in injury rates
Union Pacific	San Antonio Service Unit	Road and Yard Crews	79% decrease in locomotive engineer decertification rate. 81% decrease in derailments.
Union Pacific	Livonia Service Unit	Yard Crews	62% decrease in yard derailment rate.

These outcomes are consistent with the results obtained by the BBS process in other industries (see Krause, 1995, Chapters 16 and 17; Krause, 1997, Chapter 13; Krause, 2005, Chapter 11).

The ultimate goal of CSA and other safety culture projects was to improve the safety culture of the railroad industry. Any culture evolves slowly over time, and the CSA project does show some progress that might continue into the future. For example, at the CAB demonstration pilot at SASU "Qualitative data indicate improvements in safety awareness, personal responsibility for safety, and safety dialogue among site personnel. These improvements appear to result directly from CSA implementation activities and to mediate the improvements in practices and relations." (Zuschlag et al., 2012, p. 133). Moreover, "Labor–management relations improve with a successful CSA implementation. There is greater trust and cooperation between workers and managers." (Zuschlag et al., 2012, p. 131), a hallmark of an improved safety culture. Similarly, at the Amtrak Chicago site, "Content analysis of the interviews and focus groups conducted at the conclusion of each phase suggested positive impacts on various aspects of organizational safety culture, especially personal safety awareness and dialogue about safety." (Ranney et al., 2014, p. 17).

CSA vs. C3RS (CONFIDENTIAL CLOSE CALL REPORTING SYSTEM)

A question often asked by the FRA Office of Safety, the railroads and unions was why have two programs as similar as CSA and C3RS? Both are proactive and both are targeted in the short-term at preventing incidents and improving safety culture. The difference is that CSA is focused farther from a potential incident; C3RS is focused closer to a potential incident. This is illustrated in Figure 9.4.

In this illustration the potential incident is running a red signal (a cardinal violation of operating rules which could result in a collision and would result in decertification of the engineer). Not calling out a red signal is far less serious than almost going through a red signal, and much less serious than running a red signal. Not calling a signal and almost going through a signal could result in running a red signal and a collision. Both CSA and C3RS are proactive in that reporting and analysis precedes a potential incident rather than occurring after an incident. Both identify conditions that could cause an incident and, thereby, act to prevent incidents. CSA tends to identify risks and hazards earlier in the hazard stream than does C3RS.

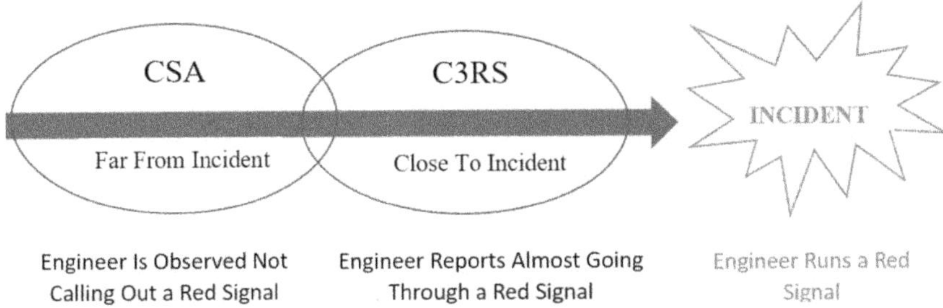

FIGURE 9.4 The difference between CSA and C3RS is distance from a potential incident.

It also provides a systematic approach for immediate coaching and feedback when those risks and hazards are first identified. The two approaches were viewed by the FRA Human Factors Research Division as complimentary processes. Table 9.2 lists key differences between CSA and C3RS. The next section discusses C3RS in detail.

CONFIDENTIAL CLOSE CALL REPORTING SYSTEMS (C3RS)

An important part of the FRA Human Factors Research Division's plan to improve safety culture in the railroad industry was the creation of a confidential close call reporting system. A C3RS program did not exist at any railroad at the time that the plan was discussed and devised. The problems associated with *ad hoc* programs, as described in the section on Safety Rules Revision, were thus avoided because the FRA initiated the program and largely controlled the process. Analyses were rigorously planned and executed to allow a thorough evaluation of the successes and failures encountered.

"In June 2002 the Federal Railroad Administration formed a broad-based Planning Committee, representing key stakeholders from industry, labor, and government. Their task was to decide how to introduce the railroad industry to the value of studying close calls as a way of improving safety.

They worked together over the next 10 months to design a workshop to meet that objective. The committee defined the agenda, the small discussion group format, and selected the speakers from other industries and railroads that have benefited from studying close calls. Each member of the Committee also briefed their own organization to encourage their attendance and support at the Workshop." (Saks et al., 2004, p. ix).

The program manager for the C3RS project was Tom Raslear. He contracted with the Volpe Center to provide support for the project. Primary project manager at Volpe was Jordan Multer. In discussions with Dr. Multer it was decided that the formation of an informal planning committee consisting of all major elements of the

---

TABLE 9.2

Key Differences Between CSA and C3RS

	CSA	C3RS
Data Input	• Employees observe each other using a checklist and give feedback • Behaviors are observed • Observee explains at-risk conditions / behaviors to peer observer	• Employees submit reports on unsafe conditions or events to a third party • Story is told
Employee's role	• Participate in peer observation; provide explanations; receive coaching • Participate in local labor problem solving • Escalate at-risk conditions to management	• Submit reports • Participate in joint labor, management, and FRA problem solving
Discipline protection	• No identifying information is collected • Observations are outside the discipline process • Waiver from FRA not required	• The third party de-identifies all reports before sharing • Employees receive protection from discipline for violations • Waiver from FRA required

railroad industry and governmental agencies concerned with railroad safety was the first step in a process to launch a close call reporting system. We were careful not to call the planning committee an advisory committee because government regulations (the Federal Advisory Committee Act, see Federal Advisory Committee Act (FACA) Management Overview | GSA) required permission from the Office of Management and Budget and the U.S. General Services Administration to form advisory committees. The process for such an endeavor is complex and time-consuming. The committee met periodically at the FRA's offices (at the time, Vermont Avenue in the District of Columbia (DC)). No costs were incurred by the FRA since committee members were either stationed in DC or traveled there frequently for their jobs. Refreshments for the group were provided by the FRA program manager.

The first task of the planning committee was to define what a "close call" was. That constituted the first problem faced by the project since FRA and Volpe members initially called near-accidents "near misses". The term "near miss" was widely used in other industries and in the human factors community. However, many committee members felt that "close call" was more appropriate in the railroad industry. The committee requested that a white paper be written to define close calls and to provide information on the safety benefits of analyzing close calls, the barriers to such analysis, and key issues that have been learned from other organizations in establishing a close call system for safety management. This task fell to Dr. Jordan Multer (Multer, 2002a, b).

The White Paper

A close call is defined as "An opportunity to improve safety practices based on a condition or incident with a potential for more serious consequences" (Multer, 2002b, p. 115). This definition was established by consensus among the members of the planning committee. This definition is broader than one that is commonly used in safety management as an event that could have resulted in personal injury, property damage or environmental damage. The reason for the use of the broader definition is that FRA has a threshold for reportable incidents which is adjusted for inflation (see Chapter 3, Data Issues). Incidents that fall below that threshold can still provide valuable information concerning safety issues and should be included in a safety management system.

The white paper discusses the safety benefits of using close calls in a safety management program. As discussed in Chapter 3, events that are neither reportable nor accountable accidents are far more frequent than accidents (see Figure 3.7). The higher frequency of close calls allows for greater power in statistical analyses and a shorter timeframe for collecting a sufficient amount of information for analyses.[6] In an evaluation of system safety, close calls can point out a system weakness that, if left unresolved could result in a serious accident. "Small accidents may be predictive of larger accidents to come." (Multer, 2002b, p. 115).

Close call data can be used in a safety management system in various ways. In a retroactive approach, close calls are used to identify why unsafe acts occurred after safety has been compromised. This is the method commonly used by safety organizations such as the FRA to determine factors that contributed to reportable accidents and how to prevent them in the future. In a proactive approach, similar close calls are analyzed to identify trends and patterns that cause failures of the safety system.

Because there are many more close calls than reportable accidents, these trends and patterns may have great predictive power. Not only can close calls show where there are weaknesses in the safety management system, they can also be used to monitor changes in system safety over time. Such monitoring can reveal conditions that were not previously known from the analysis of reportable accidents alone. Design defects, gaps in supervision, unworkable procedures, and inadequate training are examples of hidden conditions that may interact with local circumstances to cause an accident.

Close call analysis has multiple benefits for everyone in an organization. Because employees, union officials and management are all involved in the system, safety awareness becomes an organization-wide concern. Moreover, "All groups see economic benefits in reducing costs associated with reductions in time lost from injuries, damage to railroad property, damage to the environment, and time required to move the customer's goods. Productivity improves when the railroads can more effectively schedule train and maintenance operations." (Multer, 2002b, p. 117).

Lessons learned from organizations that have close call monitoring and analysis systems in place were also discussed in the white paper. These lessons include the use of a third party to collect, store and analyze the data so as to encourage reporting close calls by building and maintaining trust. The third party is required to maintain the confidentiality of the data to protect the identity of the person making the report. The system must also ensure that individuals and organizations have limited protection from liability and enforcement.

A successful close call system also engages frontline staff in the design of the system. This includes all stakeholders: labor, unions and management. The system must be accepted by a broad segment of the community. All stakeholders must feel that they have ownership of the system.

The system should be structured to organize and analyze the data. This will ensure that information can be easily retrieved and used in a model of accidents. The information that is thereby generated should be provided as feedback from the system to all levels of the organization. This ensures that other locations can benefit from trends that are detected to remedy similar system weaknesses. Furthermore, feedback provides evidence at an organizational level that the system has value and at an individual level that reports are taken seriously and acted upon.

There are seven steps in the management of a close call system, and each step has barriers that must be overcome to ensure success of the system. The steps are:

Identification
Disclosure
Distribution
Analysis
Solution identification
Dissemination to implementers
Resolution (Multer, 2002a, p. 5).

The identification of a close call is critical. This determines what types of events and behaviors will be documented for collection and analysis. A very broad concept of close calls might be more inclusive of relevant safety information but could also weaken the system with reports that are not safety relevant. A narrower concept might miss truly relevant incidents. How this is resolved also depends on who sets the criterion for which reports are included in the system.

Disclosure is a significant problem for railroad close call systems. There is a long history of distrust between labor and management in the railroad industry which mitigates against disclosure. For labor, the disclosure of a close call could result in disciplinary action, especially if an operating rule was violated. It could also result in a fine if FRA regulations were violated. For the organization, information about close calls could be used to establish liability in a FELA legal action (see Chapter 2), could result in fines under regulatory action, or become the basis of new regulations. The railroad industry has an adversarial history which must be overcome by establishing trust. The use of a third party, as noted above, to ensure confidentiality is a primary solution to this problem.

Distribution involves the process of data entry into a system that captures all the information that is pertinent to the incident so that it can be analyzed effectively. The collection and storage of the data is intertwined with the problems of disclosure. Whoever collects and stores the data must be able to ensure the confidentiality of the data. This is a thorny legal problem because civil and criminal proceedings can require the public disclosure of information in the process of discovery. So involving an entity that is capable of protecting close call data from disclosure is essential.

Analysis is a multilevel endeavor. At the individual close call level, an entry to the system must be able to be sorted into meaningful categories and prioritized according to the severity of the apparent safety risk. This is also true at the site-specific level, the organizational level, and the system level. Analysis tools are required that can process the data at all of these stages and produce useful information about safety risks. Different individuals or groups may be involved in the analysis process at these different levels.

Solution identification is also a multilevel process. A risk may be identified as unique to a site or have wider importance at the organizational or system level. Again, the process may be handled by different parties at each level because knowledge of site-specific conditions may not be available at the organization or system level, and vice-versa.

Suitable solutions need to be disseminated to the individuals who will implement the solution. This, again, may be a multilevel process. Site-specific solutions may require briefings for frontline employees about changes in operations, while changes in rules or supervision practices may be required at the organization level.

Finally, information must be gathered concerning the resolution of the problem that was addressed by the solution. Was the solution effective? Feedback must also be provided to the individual(s) who disclosed the close call. This reinforces the perception of the system as a positive safety process and encourages its use by others. An analysis system is also required to track the relationship between categories of close calls, putative solutions, and the resolution of the problem.

The white paper also touted the beneficial outcomes of close call systems in various organizations. This included:

building and maintaining trust between labor, management and regulators
engaging frontline staff in the design of the system
structuring the system so that information can be easily organized and analyzed
providing continuous feedback to all levels of the organization.

The white paper was used by committee members to solicit support for the development of a close call system in the railroad industry.

The next step taken by the committee was a recommendation that FRA sponsor a workshop on close calls. "The purpose of the workshop was to educate the railroad industry on the benefits of understanding close call events and the challenges to the implementation and success of a close call reporting system. The workshop provided a forum for participants to discuss issues of concern to the railroad industry." (Saks et al., 2004, p. i). The committee defined the agenda, selected speakers from other industries and railroads that had used close call systems and worked within their own organization to encourage attendance at the workshop.

The *Human Factors Workshop: Improving Railroad Safety Through Understanding Close Calls* was held on April 23–24, 2003 in Baltimore, MD. A complete documentation of the proceedings is contained in Saks et al. (2004). The major outcome of the workshop was the general consensus of the participants that a pilot close call project should be initiated to allow the industry to assess the usability and effectiveness of a close call system. To this end, the workshop participants requested that the planning committee "...*prepare an executive briefing* to inform leaders of all stakeholders in the railroads, regulatory agencies, and labor unions about close calls and obtain their buy-in." (Saks et al., 2004, p. xii), and continue to meet to support the development of the pilot project.

THE MODEL MEMORANDUM OF UNDERSTANDING (MOU)

An initial step in launching a pilot close call project was the development of a detailed description of the responsibilities of each of the parties to a project (FRA, labor unions, carrier, third party confidential reporting system provider), the scope of the project, and operating plan for the system. The Model MOU was formally signed by Acting FRA Administrator Robert D. Jamison and the members of the planning committee (Tom Raslear, Jim Remines, John Grundmann, Mike Coplen, Tim DePaepe, Stephen J. Klejst, Carolina Mirabal, Jeff Moller, Jordan Multer, Miriam Kloeppel, Scott Kaye, James Stem, Bob Harvey, Jo Strang, Jane Saks) on March 30, 2005 (Federal Railroad Administration, 2005).

The Model MOU defined the purpose of the project as an effort "...to gain full agreement from all parties to cooperate in the C^{3}RS Demonstration Project." (Federal Railroad Administration, 2005, p. 3). Individual railroad properties were not committed to participation in the project by the Model MOU. Rather, implementation of the project was to be achieved through separate implementing memoranda of understanding (IMOUs) between the FRA, a carrier, and a labor union at a specific site. The goals of the project were

1. "To monitor the frequency of known failure modes (existing risks to safety);
2. To learn about new failure modes (new risks to safety);
3. To maintain alertness to the risks inherent in railroad operations; and
4. To enable carriers, labor organizations and FRA to identify safety issues that require corrective action." (Federal Railroad Administration, 2005, p. 6)

The six primary functions of the project were to accept close call reports that meet the specific requirements of the MOU; store these reports as confidential information; analyze the reports; disseminate information on trends and risks; track carriers' corrective actions to evaluate the project's safety impact; and identify ways to improve the reporting system. Consequently, the Model MOU identified four key elements of the system: (1) it should be focused on impediments to safety; (2) it should be confidential and (3) voluntary; (4) it should provide participants with protection from discipline and decertification (employees) and from legal action (carriers).

The Model MOU identified the key stakeholders in the project as FRA Human Factors R&D Program; FRA Office of Safety; BTS; the Volpe Center; the carriers; the labor organizations; and the Peer Review Team (PRT), an expert team comprised of the key stakeholder representatives from FRA, the carriers and representatives from relevant labor organizations. (Federal Railroad Administration, 2005, p. 8)

The Bureau of Transportation Statistics (BTS) was designated under the Model MOU as the organization responsible for assuring the confidentiality of the submitted close call reports because it had the ability to protect the confidentiality of close call reports and their analysis through its own authorization and the Confidential Information Protection and Statistical Efficiency Act of 2002 (CIPSEA).[7] Under the Model MOU BTS had the responsibility to protect the identity of individuals submitting close call report, the report itself, the names of other employees named in the report, the carrier, and any details about locations, unique equipment, etc., that might be used to identify those involved in the close call. As noted in Section 5.3 of the Model MOU, BTS was responsible for 7 of 12 steps in the close call report process (see also Bureau of Transportation Statistics, 2005, 2007). In 2010 this role was assigned to NASA (National Academies of Sciences, Engineering, and Medicine, 2020).

The Peer Review Team (PRT). The PRT was composed of two representatives from carrier management, the affected labor organizations, FRA, and one representative from BTS. The PRT had the responsibility to

a. Analyze each close call report (after the identifying information has been removed) and root causes;
b. Analyze summarized data from multiple reports;
c. Identify new sources of emerging trends & new types of safety-critical risks;
d. Assess the association between emerging patterns or trends in close calls, relate those to corrective actions to be taken by the carriers, and advise on implementation;
e. Review and discuss a summary report comprised of the individual close call reports generated from the Close Call Reporting System, emerging trends, identified root causes and suggested corrective actions;
f. Distribute report to participating railroads and FRA giving feedback on close calls, emerging trends and newly identified risks, which were provided by BTS to the PRT; and
g. Review and discuss all reports prior to their distribution. [and] Track corrective actions taken in response to close call events. (Federal Railroad Administration, 2005, p. 9).

The Volpe Center shared responsibility with BTS for project planning and management of the reporting system (p. 16). In addition, the Volpe Center was responsible for program evaluation. This entailed the collection of baseline measures of safety and reporting culture for each participating railroad, the measurement of risk reduction during the project using baseline safety measures, the improvement of the implementation of the project by providing feedback to participants, and reports on the project at baseline, midterm and end of project (p. 17). This was considered a major component of the project since it was essential to be able to demonstrate that the project was able to reduce the risk of accidents and was cost effective.

The project at individual carrier participants was planned to operate for 5 years (p. 7). Recruiting individual carrier sites and conducting the necessary training proved to be time consuming and required multiple program evaluations to be simultaneously conducted. The Union Pacific North Platte Service Unit was the first site to participate (Federal Railroad Administration, 2006), followed by Canadian Pacific (Federal Railroad Administration, 2007b), New Jersey Transit (Federal Railroad Administration, 2009c, 2014e), Amtrak (Federal Railroad Administration, 2010, 2013a, 2014a,b,c), Strasburg Railroad Company (Federal Railroad Administration, 2013b), and Metro-North Railroad (Federal Railroad Administration, 2014d).[8]

THE OUTCOME OF THE PROJECT

FRA established and funded a dedicated program evaluation of the C3RS project. This included baseline, midterm, and final formative, summative, and sustainability evaluations (Mandler, 2014[9]) at participating carriers. Consequently, there are numerous reports documenting the program's safety outcome and lessons learned concerning implementation of the program.

Federal Railroad Administration (2008c) shows a simplified logic model for the project (see Figure 9.5). A logic model is often formulated as part of a program evaluation to describe a project's theory and assumptions concerning the activities and

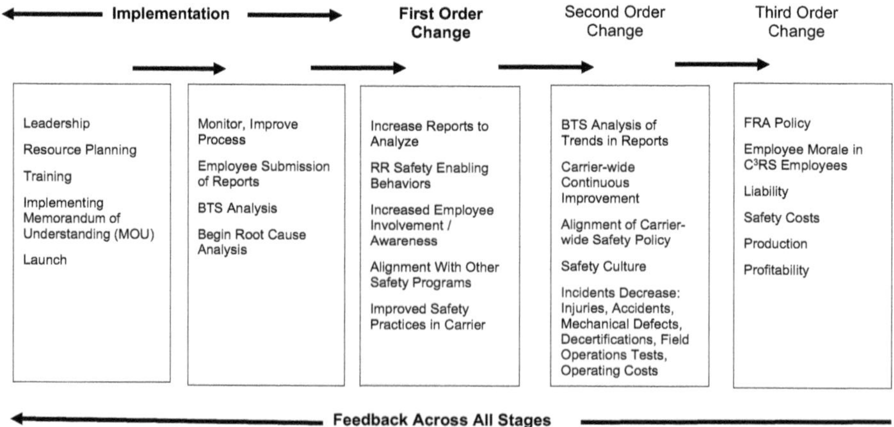

FIGURE 9.5   Logic model for C3RS project (FRA, 2008c).

expected outcomes of the project. In the case of C3RS, implementation is detailed as the first two stages of the project. Feedback across all of the stages allows implementation to be revisited if problems in later stages are discovered to be rooted in implementation. This is the formative evaluation, and its most important aspect is the monitoring of the project with the goal of improving its processes to ensure success. Success is defined by the three stages of summative evaluation: first, second, and third order changes that are qualitatively and quantitatively measurable. For instance, an early conceptualization of the carrier PRT Support Team was that only the transportation department would be involved in implementing corrective actions. However, it was found that effective implementation of corrective actions often required involvement of other departments, such as engineering, maintenance-of-way, and mechanical departments (Federal Railroad Administration, 2012a). Hence, the program evaluation team recommended that the PRT Support Team should include senior managers in multiple departments. Other interim reports found that the project faced challenges regarding the functioning of the PRT. This included setting priorities in the analysis of cases, the coordination of monitoring and implementing corrective actions with the PRT Support Team, and the cost of the project (Federal Railroad Administration, 2014f). A midterm analysis (Federal Railroad Administration, 2014g) found that carriers improved communications between the PRT and the Support Team, that the PRT began to use trend analysis on the reports and that the PRT and Support Team were tracking implementation of corrective actions.

A complete and comprehensive logic model of the project baseline can be found in Ranney et al. (2015, Figures 8, 12, 13, 14). The baseline report includes four carriers: Union Pacific, Canadian Pacific, Amtrak and New Jersey Transit. The report features a diagram that shows how the evaluation will make meaningful comparisons across sites and over time (Figure 7).

Details the quantitative and qualitative measures used in the evaluation.

Includes a diagram (logic model) that shows the relationship among the parts and expected outcomes of C3RS (Figure 6, p. 1).

In addition to this descriptive information, the baseline report contains data about perceived safety culture in labor and management, knowledge and beliefs about C3RS, and willingness to report close calls. Management rated organizational and managerial safety culture more positively than labor, while labor rated coworker safety culture more positively than organizational and managerial safety culture. Management and labor rated coworker safety culture high, indicating agreement that labor behaved safely. Regarding knowledge and beliefs about C3RS, labor had a higher awareness of conditions that might lead to an accident than managers. Labor and management agreed that changes in systems, individual behaviors and knowledge of root causes could prevent accidents. Managers were also more positive than labor that the project would be able to implement changes to improve safety. Between 60% and 82% of labor indicated that they would be willing to report close calls (pp. 34–37).

The final report on formative evaluation (Ranney et al., 2019) addressed three issues: (1) what conditions were necessary to implement C3RS as planned in the railroad industry; (2) what was the impact of C3RS on safety and safety culture; (3) what would help to sustain C3RS when the demonstration ended. The evaluation started in 2006 and ended in August 2017. Each of four demonstration sites

(Union Pacific (UP) North Platte Service Area, Canadian Pacific (CP) Chicago Area, New Jersey Transit (NJT), and Amtrak) were followed for five overlapping years (each site started at different times, hence the 10-year timeframe). The evaluation employed a rigorous quasi-experimental, mixed-methods-design evaluation. "Data included interviews with participants and stakeholders, trend analysis of close call reports and corrective actions, validated scales of safety culture, and 'bottom-line' impacts to safety." (Ranney et al., 2019, p. 1).

The final process observed in the four sites is shown in Figure 9.6. This differs somewhat from what was originally envisioned by the Planning Committee in that roles for several entities in the process are specified in more detail. This can be seen more clearly by comparing the simplified logic model in Figure 9.5 with Figure 9.7 from the Ranney et al. (2019) final report. The simplified logic model did not specify roles for the FRA, the Third Party, or elements of the railroad at the pilot sites. This is particularly true for the pilot sites. As the implementation of the project progressed at each site, the roles and actors became clearer and were able to be summarized in Figures 9.6 and 9.7.

What Conditions Were Necessary to Implement C3RS as Planned in the Railroad Industry?

This is an important question because an unsuccessful implementation makes an evaluation of the effectiveness of a program impossible. An unsuccessful implementation would prevent a program to succeed, and if the implementation was not evaluated a failed implementation might be confused with a failed program. Ranney et al. (2019) provide a model of the elements that contribute to a successful implementation of C3RS in Figure 9.8. The development of the "fishbone" diagram followed a robust process by Ranney et al. (pp. 49–50). The upper side of Figure 9.8 shows the responsibilities of each of the actors in the processes detailed in the logic model of

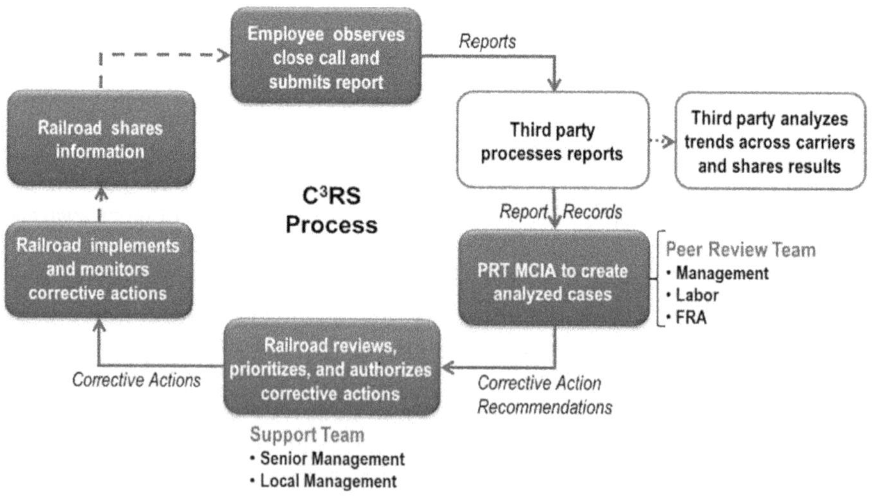

FIGURE 9.6   C3RS process (Ranney et al., 2019).

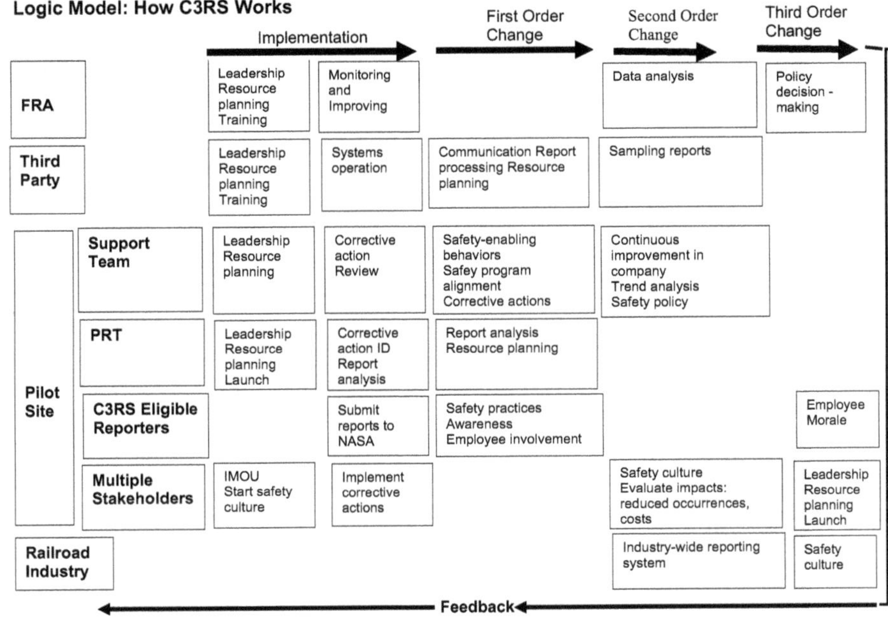

FIGURE 9.7 Final logic model for C3RS (Ranney et al., 2019).

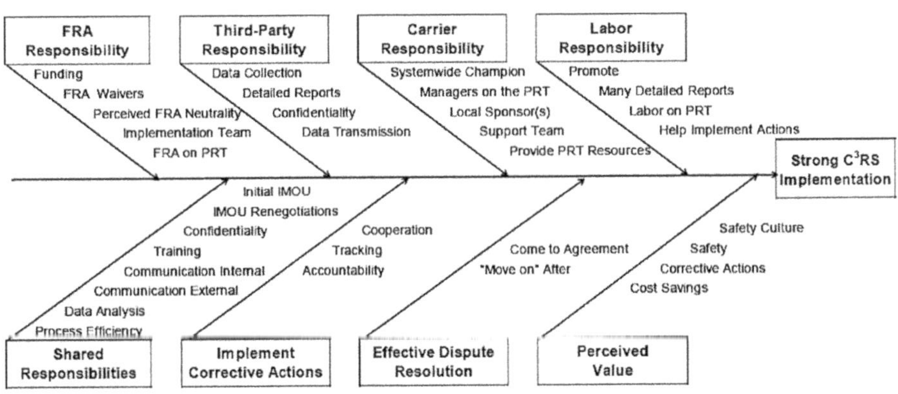

FIGURE 9.8 Implementation diagram for C3RS (Ranney et al., 2019).

Figure 9.7. These are the "causes" of "effects" shown on the lower side of Figure 9.8. Together the effects contribute to a successful implementation of C3RS.

Three of the four sites were found to have a successful C3RS implementation (see Ranney et al. Section 5.3.1 and Section 5.3.1.5, Figure 29 for details). As can be seen in Figure 9.8, there are eight factors which contribute to a successful implementation. Site 4 had "fair" or "poor ratings on six of the eight factors. Site 4 had "fair" ratings on carrier responsibility, shared responsibility, and perceived value, and "poor" ratings on labor responsibility, ability to implement corrective actions, and effective dispute resolution. By comparison the other three sites had only "fair" ratings for between one and three factors (Site 1: third party responsibility, shared responsibility, ability to implement corrective actions; Site 2: labor responsibility, effective dispute resolution; Site 3: labor responsibility).

What Was the Impact of C3RS on Safety and Safety Culture?

Given that Site 4 did not have a successful implementation, it is no surprise that there was no impact of C3RS on safety at Site 4. Figure 9.9 shows the summary impact of C3RS for each site in the demonstration project. Derailments decreased by 20%–41% at the sites with successful implementations. The decrease of 20% at Site 3 was not statistically reliable although this site had only one "fair" rating. Site 1, which had three "fair" ratings, had a 53% decrease in incident costs (also not statistically reliable). Site 2 had an 18% reduction in injuries, and Site 3 had a 39% reduction in the cost of disciplinary hearings. Taken as a whole, the results indicate that successful implementations resulted in positive impacts on four different safety concerns.

What Would Help to Sustain C3RS When the Demonstration Ended?

Sites 1 and 4 did not remain in the C3RS program when the demonstration ended. Site 4 did not have a successful implementation and did not achieve any positive safety impacts. Both Sites 1 and 4 were unable to negotiate a new IMOU to remain in the program, an issue which caused Site 1 to have a "fair" rating and for Site 4 to have a "poor" rating for Shared Responsibility. Apparently, a successful implementation and positive safety impact is not sufficient to sustain participation in C3RS. Agreement between a carrier and external shareholders (FRA, labor, third party, etc.) about program specifics are necessary to sustain participation.

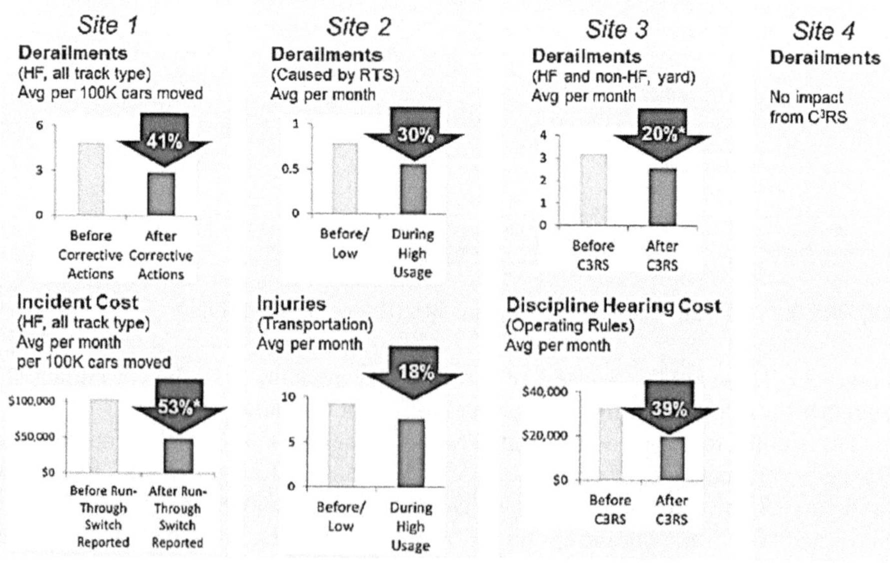

FIGURE 9.9   Summary of safety impacts of C3RS (Ranney et al., 2019).

Ranney et al. also examined sustainability of C3RS at an industry level based on interviews, project records and field notes. At the time of the report there were eight railroads and 15 labor organizations participating in the project. Of the railroads, none of the Class I and only one of the shortline carriers were participating. Ranney et al.'s general assessment of industry sustainability is shown in Figure 9.10.

Other than the absence of Class I and Shortline participation in the project, there seems to be scant data to support the Ranney et al. industry assessment of sustainability. They note that trust and cooperation between labor and the carriers are essential to agreements to implement C3RS. Moreover, they note that carriers "… supported C3RS in part because of several traumatic fatal accidents…" (Ranney et al. p. 100). Trust and cooperation between labor and the railroads is often lacking, and recent history (see Bhattacharya, 2023; Jeffries, 2023 concerning Norfolk Southern train derailment in East Palistine, OH) supports the adage that "railroad safety rules are written in blood".

In conclusion, it appears that a reporting system such as C3RS, if properly implemented, can have substantial safety impacts. However, a successful implementation and improvements in safety and safety culture alone are not sufficient for sustainability of such a program at individual sites. Moreover, a general rollout of the FRA's program appears to be hindered by the same issues that affected sustainability at two of the demonstration sites: trust and cooperation between the railroads and external shareholders, especially labor and FRA.

Other summaries of the C3RS project are available for the interested to obtain a broader view. These include Multer et al. (2013), National Academies of Sciences, Engineering, and Medicine (2020), Ranney et al. (2013), United States Government Accountability Office (2022).

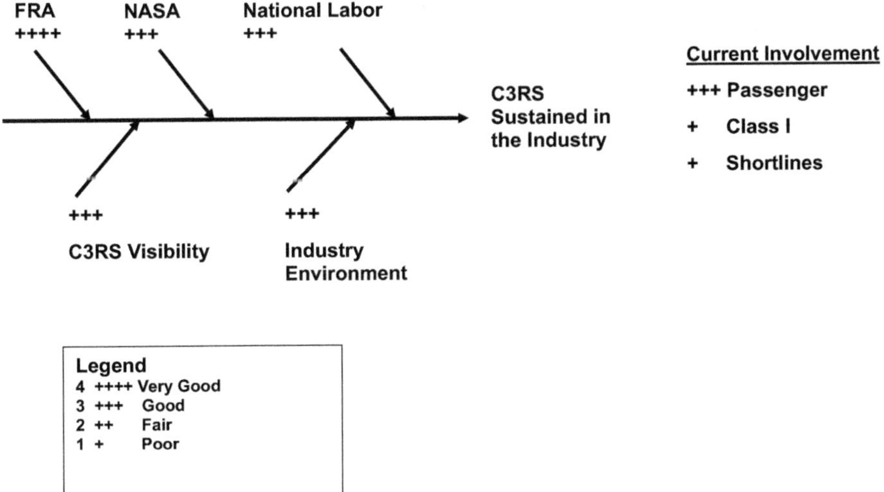

FIGURE 9.10 Sustainability of C3RS in the railroad industry by type of railroad (Ranney et al., 2019).

NOTES

1　The Human Factors Research Division incorporated a systems approach called the Sociotechnical System in the development of its research program. The details of that approach are contained in Chapters 1 and 2.
2　See Chapter 2 for a discussion of FELA
3　Delay of reinforcement is well-documented to decrease its effectiveness (Kling & Schrier, 1971, p. 677.)
4　There is an entire behavioral learning literature on the partial reinforcement effect which disagrees, somewhat, with this characterization in that a consequence (a reinforcer) which does not occur after every targeted behavior results in a stronger behavioral outcome or learning (Kling, 1971, p. 607.)
5　A negative consequence is known as punishment. Punishment will suppress behavior but not eliminate it (Church, 1969; Church, 1971).
6　In statistics power refers to the ability of a statistical procedure to detect an effect. In general, as sample size increases, the power of a statistical test also increases. If an effect is small, a sufficiently large sample size is required to detect it. The determination of a sufficient sample size is known as power analysis (Cohen, 1988).
7　By federal law (49 U.S.C. 111(i), Anonymous (2004)), data collected by BTS is not subject to the Freedom of Information Act or third party litigation. Under CIPSEA raw data cannot be used for the development of regulations. See Confidential Information Protection and Statistical Efficiency Act (2002).
8　Multiple implementing MOUs at a carrier were required because different sites or different groups of employees were involved at different times.
9　Formative evaluations occur before or during a project and are intended to improve programs by guiding planning and implementation. Summative evaluations occur after a project and are intended to prove program worth by assessing program impacts. Sustainability refers to whether a project will have longevity. Longevity is not assured by successful implementations and outcomes.

REFERENCES

Anonymous (2004). 49 U.S.C. 111(i).

Bhattacharya, A. (2023). It took the Norfolk Southern incident for major US rail companies to take a 2007 common-sense program seriously. https://www.yahoo.com/tech/took-norfolk-southern-incident-major-105900633.html.

Bureau of Transportation Statistics (2005). Memorandum of understanding between Federal Railroad Administration, U.S. Department of Transportation and Research Innovative Technology Administration, Bureau of Transportation Statistics, U.S. Department of Transportation. Unpublished document.

Bureau of Transportation Statistics (2007). *C³RS manual of operations*. Washington, DC: U.S. Department of Transportation.

Church, R. M. (1969). Response suppression. In B. A. Campbell & R. M. Church (Eds.), *Punishment and aversive behavior* (pp. 111–156). New York: Appleton-Century-Crofts.

Church, R. M. (1971). Aversive behavior. In J. W. Kling & L. A. Riggs (Eds.), *Woodworth & Schlosberg's Experimental psychology* (3rd edn, pp. 703–741). New York: Holt, Reinhart and Winston.

Cohen, J. (1988). *Statistical power analysis for the behavioral sciences* (2nd edn). Hillsdale, NJ: Lawrence Erlbaum Associates.

Confidential Information Protection and Statistical Efficiency Act (2002), 116 STAT. 2962, Public Law 107–347.

Coplen, M. (1999). *Compliance with railroad operating rules and corporate culture influences - Results of a focus group and structured interviews* (Report No. DOT/FRA/ORD-99/09). Washington, DC: U.S. Department of Transportation.

Coplen, M. (2014a). *Clear signal for action (CSA).* Unpublished document.

Coplen, M. (2014b). *Clear signal for action (CSA). Program history.* Unpublished PowerPoint TRB presentation.

Devoe, D. B., & Story, A. W. (1973). *Guidelines for writing railroad operating rules* (Report No. FRA-RT-74-1). Washington, DC: Transportation Research Board.

Federal Railroad Administration. (2003). *The impact of safety rules revisions on safety culture, incident rates, and liability claims in the U.S. railroad industry: A summary of lessons-learned* (Report No. RR03-03). Washington, DC: U.S. Department of Transportation.

Federal Railroad Administration. (2005). *Model memorandum of understanding for participating stakeholders in C3RS (Confidential Close Call Reporting System) demonstration project.* Washington, DC: U.S. Department of Transportation.

Federal Railroad Administration. (2006). *Confidential close call reporting system implementing memorandum of understanding (C3RS/IMOU) North Platte Service Unit of the Union Pacific Railroad Company.* Washington, DC: U.S. Department of Transportation.

Federal Railroad Administration. (2007a). *Behavior-based safety at Amtrak-Chicago associated with reduced injuries and costs* (Report No. RR 07-07). Washington, DC: U.S. Department of Transportation.

Federal Railroad Administration. (2007b). *Confidential close call reporting system implementing memorandum of understanding (C3RS/IMOU) of the Canadian Pacific.* Washington, DC: U.S. Department of Transportation.

Federal Railroad Administration. (2008a). *Promising evidence of impact on road safety by changing at-risk behavior process at Union Pacific* (Report No. RR 08-08). Washington, DC: U.S. Department of Transportation.

Federal Railroad Administration. (2008b). *Positive safety outcomes of clear signal for action program at Union Pacific yard operations* (Report No. RR 08-09). Washington, DC: U.S. Department of Transportation.

Federal Railroad Administration. (2008c). *Confidential close call reporting system: Preliminary evaluation findings* (Report No. RR 08-33). Washington, DC: U.S. Department of Transportation.

Federal Railroad Administration. (2009a). *Decreases in collision risk and derailments attributed to changing at-risk behavior process at Union Pacific* (Report No. RR 09-20). Washington, DC: U.S. Department of Transportation.

Federal Railroad Administration. (2009b). *Safe practices, operating rule compliance, and derailment rates improve at Union Pacific yards with STEEL process – A risk reduction approach to safety* (Report No. RR 09-08). Washington, DC: U.S. Department of Transportation.

Federal Railroad Administration. (2009c). *Confidential close call reporting system implementing memorandum of understanding (C3RS/IMOU) for New Jersey Transit.* Washington, DC: U.S. Department of Transportation.

Federal Railroad Administration. (2010). *Confidential close call reporting system implementing memorandum of understanding (C3RS/IMOU) for Amtrak.* Washington, DC: U.S. Department of Transportation.

Federal Railroad Administration. (2012a). *Senior cross-functional support — essential for implementing corrective actions at C3RS sites* (Report No. RR 12-09). Washington, DC: U.S. Department of Transportation.

Federal Railroad Administration. (2013a). *Confidential close call reporting system implementing memorandum of understanding (C3RS/IMOU) for AMTRAK.* Washington, DC: U.S. Department of Transportation.

Federal Railroad Administration. (2013b). *Confidential close call reporting system implementing memorandum of understanding (C3RS/IMOU) Strasburg Railroad Company.* Washington, DC: U.S. Department of Transportation.

Federal Railroad Administration. (2014a). *Confidential close call reporting system implementing memorandum of understanding (C3RS/IMOU). Amtrak.* Washington, DC: U.S. Department of Transportation.

Federal Railroad Administration. (2014b). *Confidential close call reporting system implementing memorandum of understanding (C3RS/IMOU) Amtrak* [Hialeah Mechanical Shop]. Washington, DC: U.S. Department of Transportation.

Federal Railroad Administration. (2014c). *Confidential close call reporting system implementing memorandum of understanding (C3RS/IMOU) Amtrak* [all Amtrak owned or operated Mechanical Shops]. Washington, DC: U.S. Department of Transportation.

Federal Railroad Administration. (2014d). *Confidential close call reporting system implementing memorandum of understanding (C3RS/IMOU) Metro-North Railroad.* Washington, DC: U.S. Department of Transportation.

Federal Railroad Administration. (2014e). *Confidential close call reporting system implementing memorandum of understanding (C3RS/IMOU) for New Jersey Transit* [Final]. Washington, DC: U.S. Department of Transportation.

Federal Railroad Administration. (2014f). *Update from C3RS lessons learned team: Four demonstration pilots* (Report No. RR 14-17). Washington, DC: U.S. Department of Transportation.

Federal Railroad Administration. (2014g). *Update from C3RS lessons learned team: Safety culture and trend analysis* (Report No. RR 14-18). Washington, DC: U.S. Department of Transportation.

Keller, F. S. & Schoenfeld, W. N. (1950). *Principles of psychology.* New York: Appleton-Century-Crofts.

Kling, J. W. (1971). Learning. Introductory survey. In J. W. Kling & L. A. Riggs (Eds.), *Woodworth & Schlosberg's experimental psychology* (3rd edn, pp. 551–613). New York: Holt, Reinhart and Winston.

Kling, J. W. & Schrier, A. M. (1971). Positive reinforcement. In J. W. Kling & L. A. Riggs (Eds.), *Woodworth & Schlosberg's experimental psychology* (3rd edn, pp. 613–702). New York: Holt, Reinhart and Winston.

Krause, T. R. (1995). *Employee-driven systems for safe behavior.* New York: Van Nostrand Reinhold.

Krause, T. R. (1996). *The behavior-based safety process. Managing involvement for a injury-free culture.* New York: Van Nostrand Reinhold.

Krause, T. R. (2002). *An overview of Behavior Based Safety.* Unpublished PowerPoint presentation.

Krause, T. R. (2005). *Leading with safety.* Hoboken, NJ: Wiley.

Krause, T. R. & Bell, K. J. (2015). *7 Insights into safety leadership.* Ojai, CA: Safety Leadership Institute.

Leveson, N. G. (1995). *Safeware. System safety and computers.* Reading, MA: Addison-Wesley.

Mandler, M. (2014). *Project to institutionalize program evaluation at Volpe. Technical center directors briefing, Volpe, The National Transportation Systems Center.* Unpublished PowerPoint presentation.

Moray, N. (1994). Error reduction as a systems problem. In M. S. Bogner (Ed.), *Human error in medicine* (pp. 67–91). Hillsdale, NJ: Lawrence Erlbaum.

Multer, J. (2002a). *Improving safety through understanding close call events.* Paper presented at the meeting of the Federal Railroad Administration Close Call Steering Committee, Washington, DC.

Multer, J. (2002b). Improving safety through understanding close calls. In J. Saks, J. Multer, & K. Blythe (Eds.), *Proceedings of the human factors workshop: Improving railroad safety through understanding close calls* (Report No. DOT/FRA/ORD-04/03, pp. 113–120). Washington, DC: U.S. Department of Transportation.

Multer, J., Ranney, J., Hile, J. & Raslear, T. (2013). Developing an effective corrective action process: Lessons learned from operating a confidential close call reporting system. In N. Dadashi, A. Scott, J. R Wilson, & A. Mills (Eds.), *Rail Human Factors: Supporting reliability, safety and cost reduction.* London: Taylor & Francis, pp. 659–669. 659. (Report No. DOT/FRA/ORD-13/12).

National Academies of Sciences, Engineering, and Medicine. (2020). *Characteristics and elements of nonpunitive employee safety reporting systems for public transportation.* Washington, DC: The National Academies Press. https://doi.org/10.17226/25852.

Ranney . J. (2003). *Safety culture.* Unpublished PowerPoint Presentation at the Annual Meeting of the Transportation Research Board, TRB Human Factors in Transportation Workshop 105.

Ranney, J. M., Davey, M., Morell, J., & Raslear, T. G. (2015). *Confidential close call reporting system (C3RS) lessons learned baseline phased report* (Report No. DOT/FRA/ORD-15/08). Washington, DC: U.S. Department of Transportation.

Ranney, J. M., Davey, M., Morell, J., Zuschlag, M., & Kidda, S. (2019). *Confidential close call reporting system (C3RS) lessons learned evaluation – final report* (Report No. DOT/FRA/RPD-19/01). Washington, DC: U.S. Department of Transportation.

Ranney, J., & Nelson, C. (2004). Impacts of participatory safety rules revision in U.S. railroad industry. An exploratory assessment. *Transportation Research Record,* 1899, 156–163.

Ranney, J., & Nelson, C. (2007). T*he impact of participatory safety rules revision on incident rates, liability claims, and safety culture in the U.S. railroad industry* (Report No. DOT/FRA/ORD-07/14). Washington, DC: U.S. Department of Transportation.

Ranney. J. M., Zuschlag, M. K., Coplen, M., & Nelson, C. (2014). *Peer-to-peer feedback on safety associated with reduced injuries and improved safety culture at Amtrak-Chicago.* Unpublished manuscript.

Ranney . J. M., Zuschlag, M. K., Morrell, J., Coplen, M. K., Multer, J., & Raslear, T. G. (2013). Evaluations of demonstration pilots produce change. Fourteen years of safety-culture improvement efforts by the Federal Railroad Administration. *TR News, 286,* 28–36.

Raslear, T. G. (2011). *Behavioral science for fatigue accident risk management.* Unpublished PowerPoint presentation at the Institutes for Behavior Resources 50th Anniversary.

Reason, J. (1997). *Managing the risks of organizational accidents.* Aldershot: Ashgate.

Reynolds, G. S. (1975). *A primer of operant conditioning.* (Rev. ed.). Glenview, IL: Scott, Foresman and Company.

Saks, J., Multer, J., & Blythe, K. (2004). *Proceedings of the human factors workshop: Improving railroad safety through understanding close calls* (Report No. DOT/FRA/ORD-04/03). Washington, DC: U.S. Department of Transportation.

Skinner, B. F. (1938). *The behavior of organisms.* New York: Appleton-Century-Crofts.

United States Government Accountability Office. (2022). *Federal Railroad Administration. Better communication of safety information could improve the close call system. Report to Congressional Committees.* (Report No. GAO-23-105287). Washington, DC: United States Government Accountability Office.

Zuschlag, M. K., Ranney, J. M., Coplen, M. K., & Harnar M. A. (2012). *Transformation of safety culture on the San Antonio Service Unit of Union Pacific Railroad* (Report No. DOT/FRA/ORD-12/16). Washington, DC: U.S. Department of Transportation.

10 Track Inspection

Track inspection is an important safety regulatory function of the FRA. Regulations governing track safety standards are contained in 49 CFR 213. The Track Safety Standards (TSS) of 49 CFR 213, "...contain requirements that FRA believes necessary to maintain safe track and a stable, viable rail network. These standards are *minimum* safety standards. The TSS define nine classes of track, Class 1 being the lowest quality track and Class 9 being the highest. The TSS also prescribe the maximum operating speed for freight and passenger operations for each class of track." (Al-Nazer et al., 2011, p. 5). The TSS contains standards for the general condition of each class of rail as well as standards for roadbed, track geometry, track structure, track appliances and track-related devices, inspection, and train operations at track classes 6 and higher. "In order to maintain a safe rail network, the TSS specifically prescribe several mandatory inspection techniques and procedures, including routine visual track inspections. The TSS also address threshold safety levels for several automated inspection systems, including internal rail flaw, track geometry, and gage restraint measurement systems (GRMS)." (Al-Nazer et al., 2011, pp. 6–7).

The specific conditions that violate TSS are numerous and comprehensive. "The goal of the inspection process is to prevent incipient failure of the railroad track and the resulting derailment potential of a rail vehicle. The current standards for the various track components in the TSS are divided into three primary sections for track classes 1–5: roadbed (Subpart B), track geometry (Subpart C), and track structure (Subpart D). Portions of Subpart G address similar requirements for track classes 6–9. Subpart B prescribes standards for the roadbed. This section briefly addresses the issues of water drainage and excessive vegetation.

Excessive vegetation presents a fire hazard and can obstruct the visibility of railroad signs and signals. Standing water can result in a softened roadbed that does not provide sufficient support for the track structure, including the rails, ties, and ballast. Subpart C prescribes standards for track geometry parameters, including track gage and cross-level, as well as vertical and lateral deviations of each rail. These components include ballast, crossties, rails, rail joints, anchors, and fasteners. Finally, Subpart D prescribes the standards for each of the critical components that comprise the track structure. The roadbed and track structure sections primarily address issues that can be evaluated visually. Track geometry is one measure of the overall condition of the track structure. Degraded track geometry may be the result of defective conditions that can be detected visually, such as those addressed in the roadbed and track structure sections.

However, degraded track geometry can also result from conditions that are not visible on the surface. For example, internally rotted wooden ties, abrasion on the bottom surface of concrete ties, or poor subgrade conditions may not be initially detectable by visible means, but if these conditions exist in excess, then degraded track geometry will result when the track is loaded by a rail vehicle. Track geometry,

DOI: 10.1201/9781003500018-14

therefore, can be thought of as a means for quantifying the overall condition of the track structure and substructure." (Al-Nazer et al., 2011, p. 8).

The railroad industry is responsible for the conduct of regularly scheduled visual track inspections, either by foot or on a hi-rail vehicle.[1] In addition, a variety of automated inspection and monitoring systems are used. The industry uses three inspection systems: internal rail inspection, track geometry inspection, and gage restraint measurement systems (GRMS).

Internal rail inspection is used to detect rail flaws which develop inside the rail which are not detectable from a visual inspection of the rail surface. Ultrasonic techniques are predominantly used to detect internal rail defects, but other methods are emerging and are likely to be used in the future.

Track geometry measurement systems (TGMS) measure the "…vertical and lateral deviations of each of the rails as well as the gage and cross-level measurements, which are, respectively, the horizontal and vertical relationships between the two rail heads. Maintaining proper gage and cross-level is essential, as gage or cross-level that is too large or small could result in a derailment. The vertical and lateral deviations of each of the rails are also important, as poor vertical and lateral alignment can result in high dynamic forces and, potentially, a derailment." (Al-Nazer et al., 2011, pp. 13–14). In the past track geometry measurements were conducted manually, but currently automated TGMS is the standard method.

"GRMS measures the ability of the track structure to maintain its gage under a constant, vehicle-applied gage-spreading load; by doing so, it quantifies the strength of a track's gage. GRMS vehicles include a third axle that applies a constant lateral, as well as vertical, load on each of the rails. GRMS typically measures the gage at two points: the first near the third axle and the second several feet away at an unloaded point. Based on the applied loads, the gage strength is then quantified by comparing the loaded and unloaded gage measurements. Measurement of gage strength, through the use of GRMS technology, allows railroads to effectively focus their tie and fastener replacement programs." (Al-Nazer et al., 2011, p. 14).

The routine visual inspections which are required by the TSS are detailed in 49 CFR 213.233 for track classes 1–5, and in 49 CFR 213.365 for track classes 6–9. Particularly pertinent with regard to the Track Inspection Time Study (Al-Nazer et al., 2011), 49 CFR 213.233 (b) specifies that "If a vehicle is used for visual inspection, the speed of the vehicle may not be more than 5 miles per hour when passing over track crossings and turnouts, otherwise, the inspection vehicle speed shall be at the sole discretion of the inspector… (p. 16)." As noted below, Congress mandated that FRA determine whether the speed of inspection vehicles under 49 CFR 213.233 (b) allowed for adequate track inspection. This involved a theoretical model of defect detection accuracy by track inspectors, a determination of current practice regarding inspection vehicle speed and territory in a track inspector survey, and documentation of track defects detected by inspectors.

TRACK INSPECTION TIME STUDY

The Rail Safety Improvement Act of 2008, Section 403(A)(4) (RSIA) required a report to Congress based on "…a study to determine whether…the speed at which railroad track inspection vehicles operate and the scope of the territory they generally cover allow for proper inspection of the track and whether such speed and appropriate scope should be regulated by the Secretary." This part of RSIA basically requires a determination of the maximum speed at which a visual track inspection can occur to detect track defects. An ideal observer model was constructed for visual track inspection. The ideal observer model provides a theoretical upper limit, based on previously compiled empirical data, on how well the best possible observer can perform. The primary purpose of the ideal observer model is to have a baseline against which current inspection speeds reported in the survey can be compared.

THEORY OF THE IDEAL OBSERVER

One way to determine the maximum speed at which a visual track inspection can occur is to construct a theory of the ideal observer. The ideal observer provides an upper limit on how well the best possible observer can perform by making optimal use of the information available from the stimulus (Wickens, 2002). In the present case, the maximum speed will depend critically upon human visual and cognitive capabilities and limitations. To the extent that all relevant human capabilities and limitations are known and are accurately captured, the ideal observer provides a benchmark against which actual performance and practices can be compared. In the present case, the ideal observer can serve as a comparison to practices reported in surveys, interviews and focus groups regarding the speed at which visual inspections are routinely performed.

Maximum speed in miles per hour (mph) can be defined as

$$\text{mph} = r/t \tag{10.1}$$

where r is the distance from the observer to the object to be detected and t is the exposure time (the amount of time required by the observed to detect the object with some degree of accuracy).

There are several aspects to this from a Human Factors point of view that must be considered:

1. Visual acuity and detection accuracy
2. Visual search time
3. Attention (vigilance, distraction, fatigue).

1. Visual acuity is the limit in the ability to resolve detail (Boyce, 1997, p. 872). Visual acuity is generally specified in terms of the visual angle (θ) between the observer's eye and the largest dimension (l) of the object being viewed (Figure 10.1).

If θ is small, then

$$\theta = \frac{57.3l}{r} \tag{10.2}$$

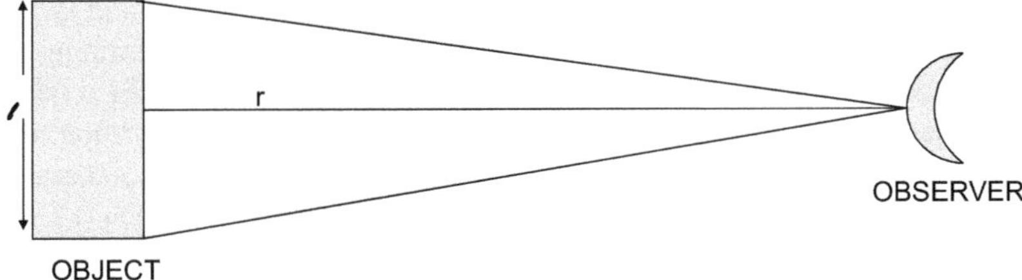

FIGURE 10.1 Relationship between size of the largest dimension (l) of the object to be detected and the distance (r) from the object to the observer. This is used to define the visual angle (θ) as noted in the text.

in degrees (Graham, 1951, p. 872), where r is again the distance of the object from the observer.

Visual acuity (θ) is a function of light intensity and exposure time (Riggs, 1972, p. 300).

Light Intensity. If we assume that track inspections take place during daytime and under good visibility conditions, visual acuity is at a plateau of approximately 0.5 min of arc (Bartley, 1951, p. 958). This, however, is a threshold value (50% detection). Visual acuity in the psychophysics literature is generally expressed as the smallest visual angle at which detection can occur with 50% accuracy. Detection accuracy is best thought about as a probability: accuracy can approach zero and 100%, but never reaches either limit. A detection accuracy of 50% (probability = 0.5) is considered the detection threshold in classical psychophysics. Intuitively, a much higher detection accuracy for safety critical tasks such as track inspection, would be desired. A rule of thumb suggested by Boyce (1997, p. 873) is that the visual angle needs to be four times bigger than threshold for quick resolution without affecting visual performance. However, a 4-fold increase in visual angle from threshold only results in 59.5% correct detection (see Appendix 12.A or Egan, 1975, p. 81 for computational details and definitions of S/N and $d'_{Y/N}$). A value above 90% would seem better suited for a safety critical task. Table 10.1 shows the visual angles for 50%, 59.5%, 90%, 95%, and 99% correct detection.

Exposure Time. The Bunsen-Roscoe Law of photochemistry is obeyed if the eye is exposed to light for short periods of time, such that the amount of energy (E)

TABLE 10.1

Percent Correct Detection, $d'_{Y/N}$, S/N, and Visual Angle

Percent Correct Detection	$d'_{Y/N}$	S/N	Visual Angle, θ (min arc)
50	0.085	1.0	0.5
59.5	0.34	4.0	2.0
90	1.82	21.4	10.71
95	2.33	27.45	13.73
99	3.29	38.77	19.38

needed to detect an object is constant and E is the product of time (t) and luminance (L). This law does not apply above a critical duration, t_c. Above t_c luminance alone determines acuity. Values of t_c range from 0.01 to over 0.2 seconds. So, to set exposure time, according to Riggs (1972, p. 304), "The duration of the best 'look' must be greater than t_c, but it need not be longer than one or two tenths of second at daylight levels of luminance." Accordingly, we set minimum exposure time (t) at 0.2 s.

It is now possible to specify the maximum speed at which objects of various sizes can be detected with a particular degree of accuracy by combining Equations (10.1) and (10.2):

$$\text{mph} = 57.3l/t\ \theta \qquad (10.3)$$

This is shown in Figure 10.2 for 50%, 59.9%, 90%, 95%, and 99% correct detection rates.

The size of the smallest object that must be detected would set the limit on track inspection speed, considering only visual acuity and desired detection accuracy. The smallest object in Figure 10.2 is 0.1 inches and can be detected with 99% detection rate at 5 mph. The same object would be detected 90% of the time at approximately 9 mph. At threshold, the object would be detected 50% of the time at a speed of 195 mph.

2. Visual search time has been extensively studied and an excellent summary of that work can be found in Luce (1986, p. 428). The average time required to search for an item is given by

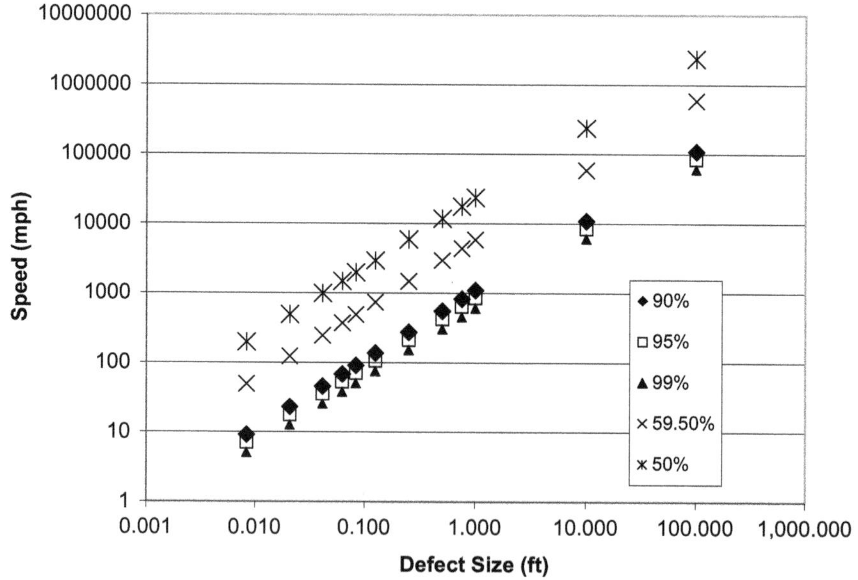

FIGURE 10.2 Relationship between maximum inspection speed, defect size and detection accuracy.

$$S = kM + r_0 \tag{10.4}$$

where k is the mean time per item, M is the number of items and r_0 is the residual time. The values of k and r_0 are 40 and 400 msec, respectively. The search time would include the exposure time in order to estimate total time available to traverse the distance to the object. In the case where there is only one object to be found in an otherwise blank field (an unrealistic scenario, but one which sets a lower boundary), the search and exposure time would be 440 msec. If we substitute S in Equation (10.4) for t in Equation (10.3), the maximum speed for track inspections is now

$$\text{mph} = 57.3l/S\theta \tag{10.5}$$

Figures 10.3–10.5 show this for 90%, 95% and 99% accuracy.

It is easy to see that as the number of search objects increases, the time to search increases and so the speed of the inspection must decrease. Again, using the smallest object (0.1 inch) as the example, maximum speed with 90% accuracy is approximately 4, 3, 1.75, 0.6 and 0.2 mph with 1, 4, 16, 64 and 256 objects to search. At 95% accuracy, maximum speed is approximately 3, 2.5, 1.4, 0.5 and 0.13 mph with 1, 4, 16, 64 and 256 objects to search. At 99% accuracy, maximum speed is approximately 2.3, 1.8, 1, 0.3 and 0.1 mph with 1, 4, 16, 64 and 256 objects to search.

3. Attention (vigilance, distraction, fatigue), is a complex issue to model in a task like track inspection. It should be noted that this is a broad subject with many components that may require separate consideration. For example, if one is fatigued, it may be easier to be distracted and harder to maintain vigilance. However, for purposes of

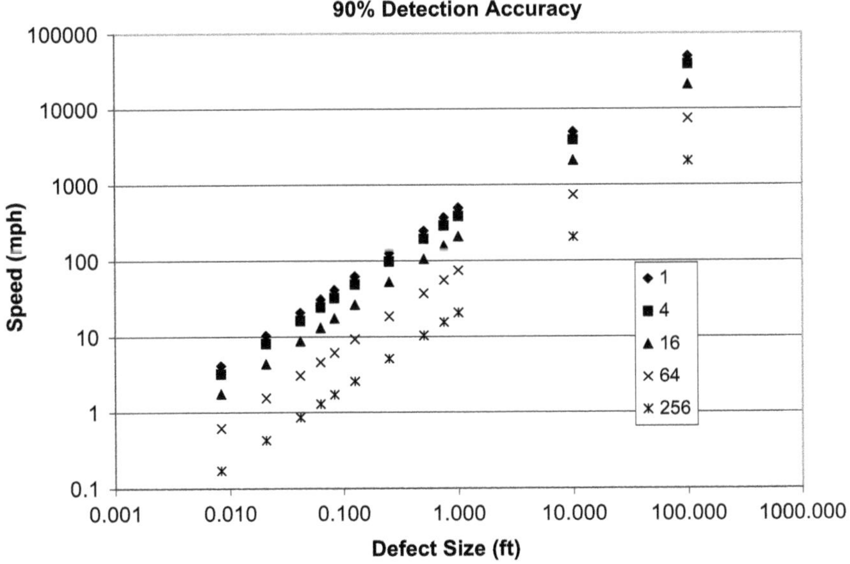

FIGURE 10.3 Relationship between maximum inspection speed, defect size, and number of search objects for 90% detection accuracy.

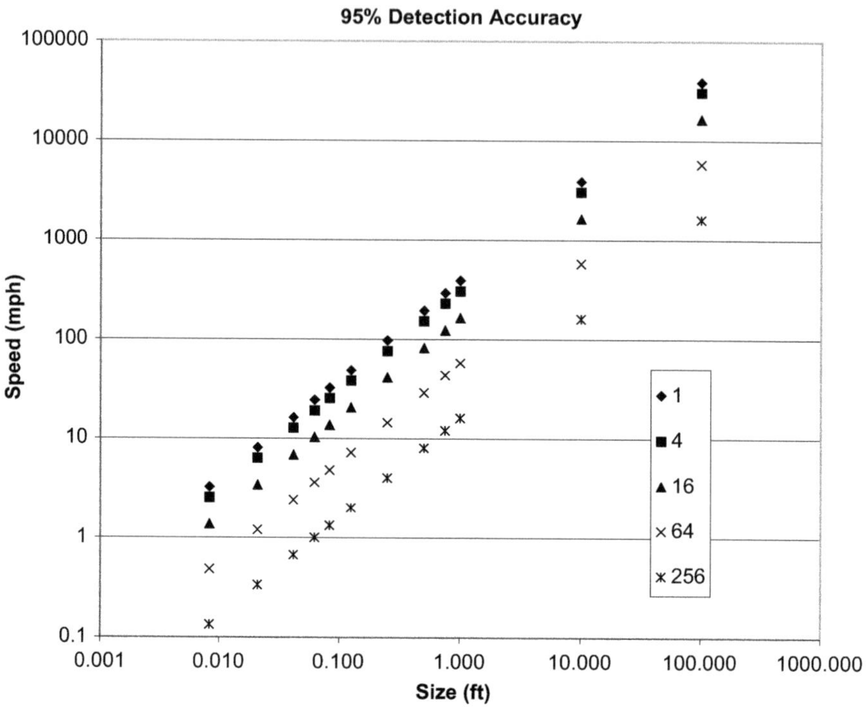

FIGURE 10.4   Relationship between maximum inspection speed, defect size and number of search objects for 95% detection accuracy.

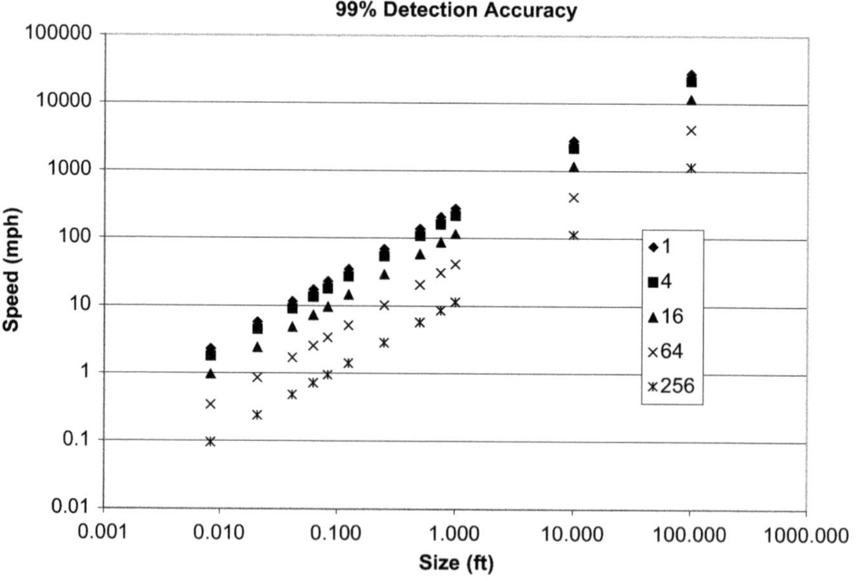

FIGURE 10.5   Relationship between maximum inspection speed, defect size and number of search objects for 99% detection accuracy.

this discussion, we will simplify and consider track inspection task to be primarily a vigilance task. This is consistent with the need to search for defects among multiple distracters (search objects) while visual information is flowing by at the speed of inspection.

The original studies on vigilance were performed by Mackworth (1950), and subsequent studies have found the basic phenomenon to be robust. According to Boff and Lincoln (1988, Vol II, p. 1504): "The correct detection of signals (hit rate) in most simple vigilance tasks shows a decrement over time. When hit rate is averaged for blocks of time (e.g., 30 min) in a 2-h task, the vigilance decrement is greatest between the first and second blocks."

According to Luce (1986, p. 176), at low signal rates correct detections in the Mackworth vigilance experiments fell from 85% in the first half hour to 73% in the rest of the watch. This corresponds to a change in $d'_{Y/N}$ from 1.47 to 0.86 or 0.61 units. Thus, to maintain performance at a desired level of percent correct detection it would be necessary to increase the visual angle sufficiently to compensate for this amount of vigilance decline after the first half hour.

Figures 10.6–10.8 show the relationship between maximum inspection speed, defect size and number of search objects for vigilance decrement adjusted detection accuracy. Again, using the smallest object (0.1 inch) as the example, maximum speed with 90% accuracy is approximately 3, 2.4, 1.3, 0.46 and 0.13 mph with 1, 4, 16, 64 and 256 objects to search. At 95% accuracy, maximum speed is approximately 2.6, 2, 1.1, 0.38 and 0.11 mph with 1, 4, 16, 64 and 256 objects to search. At 99% accuracy, maximum speed is approximately 1.9, 1.5, 0.83, 0.29 and 0.08 mph with 1, 4, 16, 64 and 256 objects to search.

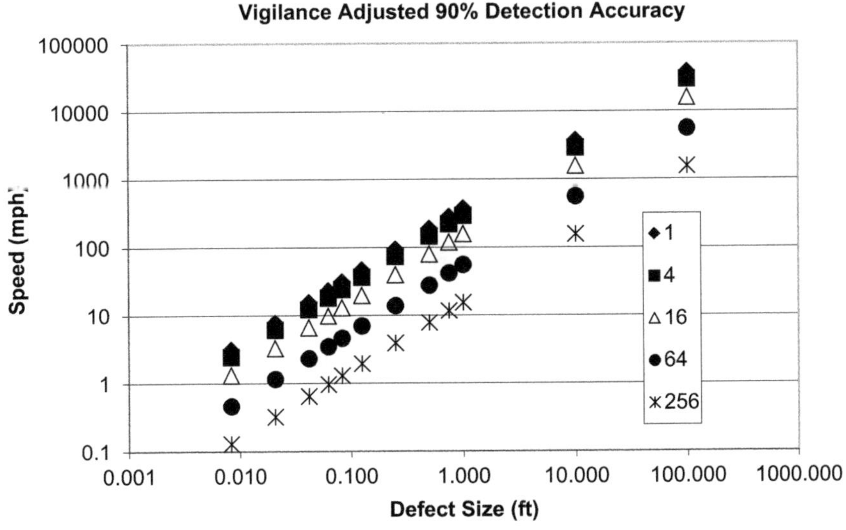

FIGURE 10.6 Relationship between visual inspection speed, defect size, and number of search objects for vigilance adjusted 90% correct detection accuracy.

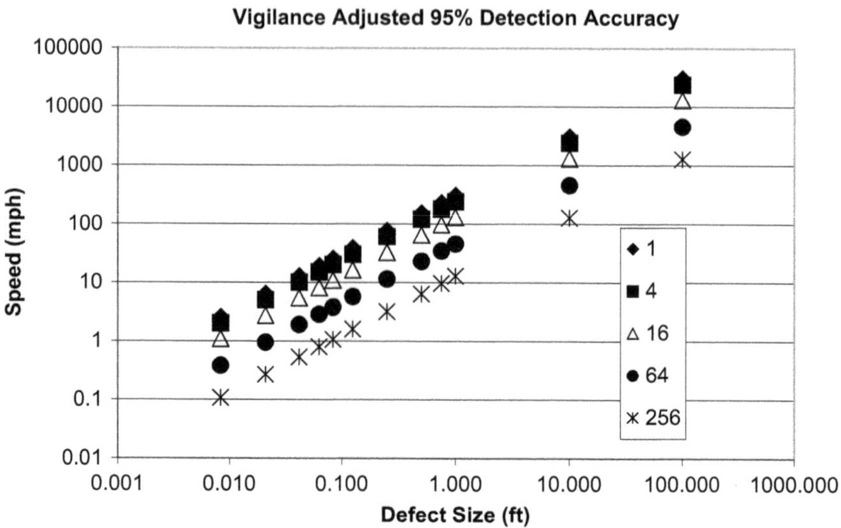

FIGURE 10.7 Relationship between visual inspection speed, defect size, and number of search objects for vigilance adjusted 95% correct detection accuracy.

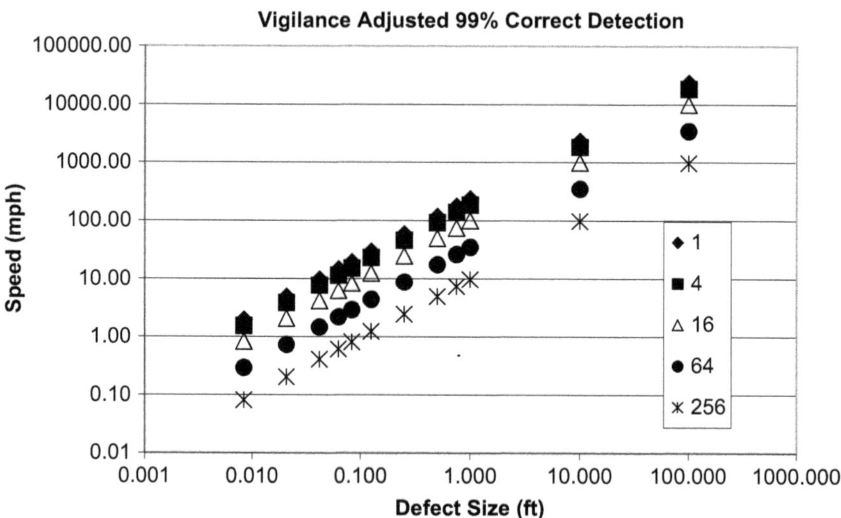

FIGURE 10.8 Relationship between visual inspection speed, defect size, and number of search objects for vigilance adjusted 99% correct detection accuracy.

TRACK INSPECTOR SURVEY

One of the factors considered in the ideal observer model that could affect detection accuracy was fatigue. Work schedules and work hours are known to directly relate to fatigue (see Chapter 7. Fatigue and Hours of Service Regulations). The track inspector survey found that "Nearly 85% of respondents indicated that they had an

8 h workday with 10% indicating that their workday was 10 h. The remaining 5% had scheduled workdays of either 8.5, 9, 9.5, 12, or 14 h." (Al-Nazer et al., 2011, p. 27). More than 50% of respondents worked overtime on 1–10 days per month. Moreover, over 50% of respondents indicated that they worked on 1–3 rest days. The survey revealed that inspectors spend an average of 5 h inspecting track and 1.9 h doing repairs. Although a specific, formal study was not conducted of work and rest schedules for track inspectors, a work-rest study was conducted for maintenance-of-way (MOW) employees which would include track inspectors (see Gertler and Viale, 2006). In the Gertler and Viale study track inspectors would be included in the non-production MOW employees. These employees report similar work schedules to the track inspectors, and more than 60% of the non-production employees reported that they always or frequently felt alert at work (Gertler and Viale, 2006 pp. 28–29). Non-production MOW employees work mostly during daytime (Gertler et al., 2013, Figure 2, p. 16) and compared to the U.S. population get significantly more sleep (Gertler et al. Table 15, p. 37, and Figure 21, p. 39). The Gertler et al. study also used the MOW work-rest data to assess the fatigue status of MOW employees using the SAFTE fatigue model (Hursh et al., 2004). They found that fewer than 30% of MOW employees had any level of fatigue (Figure 30, p. 53). Consequently, it is clear that fatigue is not an issue for the track inspectors because of their regular daytime schedules.

The amount of track that inspectors were responsible for ranged from 10 to 160 track miles[2] with a mean of 78.9. Given the amount of time spent inspecting track, this means that the average speed of track inspection (hi-rail and on foot) is 15.78 mph. Table 3 in Al-Nazer et al. (2011) indicates that an average of 65.9 track miles per day is inspected by hi-rail only or mostly. This gives an estimated track inspection speed of 13.18 mph. 79.1 track miles are inspected by hi-rail only, yielding an inspection speed of 15.82 mph. There is a correlation between an inspector's territory size and speed of inspection of 0.5. This indicates that 25% of the variance in inspection speed is accounted for by territory size.

However, there are different types of track which, according to Figure 12 in Al-Nazer et al. (2011), are inspected at different speeds using hi-rail vehicles. Curved and tangent[3] jointed[4] track is inspected a median speed between 1 and 20 mph. Tangent CWR[5] rail is inspected at a median speed between 11 and 30 mph.

TRACK DEFECTS

A total of 350,719 defects was found and documented by FRA track inspectors between 2006 and 2009. The data indicates that the predominant defects identified during this period were related to turnout and rail joint issues. These two defects accounted for more than 40% of the total defects reported. Degraded or weak fastening is the most frequent defective condition attributed to track turnouts. "Joint bars exhibit many failure modes including center cracks near the rail end gap, cracks emanating from bolt holes, as well as potentially more benign conditions such as missing or loose bolts. Bolt-hole cracks are the cause of the majority of derailments attributed to joint bar defects." (Al-Nazer et al., 2011, p. 41).

Al-Nazer et al. (p. 39) decided that an ideal observer with 95% detection accuracy, searching for 16 items simultaneously (based on the number of different defects that railroad track inspectors routinely look for), and constrained by vigilance was a reasonable model. As shown above in Figure 10.7, the ideal observer model suggests an inspection speed of 16.3 mph if 16 variable 0.125 ft objects are being searched, constrained by vigilance. The speed of track inspection in a hi-rail vehicle ranges from 5 to 30 mph depending on the type of track being inspected. Al-Nazer et al. calculated the speed for inspections primarily by hi-rail as 13.9 mph, and this value corresponds well with the output of the ideal observer model. However, "Track inspectors are searching for various size objects, many of which are much larger than 1.5 inches, such as fouled ballast and excessive wayside vegetation. Furthermore, inspectors use auditory as well as visual signals to detect defective track conditions. For example, loose joints, rail mismatch, or other variation in the running surface often create a distinct audible noise when the hi-rail vehicle moves over them. In conclusion, the ideal observer analysis, therefore, generally suggests that current inspection speeds are adequate." (Al-Nazer et al. p. 40).

AUTOMATED vs. HUMAN TRACK INSPECTION

As technology advances in the railroad industry, there has become an increasing use of automation to enhance or replace humans in the performance of railroad operational tasks. As noted above, visual track inspection is required for track classes 1 through 5 (49 CFR 213.233) and for track classes 6 through 9 (49 CFR 213.365). In addition, the TSS requires automated track inspection with a TGMS for track classes 6 through 9. However, "All Class 1 railroads make use of geometry cars even though the great majority of the track owned by Class 1 railroads is track class 5 or lower." (Al-Nazer et al., 2017, p. 5). The use of TGMS on lower class track is not mandated by the TSS, but is assumed to enhance the safety of these track classes. However, the effectiveness of any automated system is dependent on how it is used in combination with human operators (see France-Peterson et al., 2023). Consequently, it is important to know whether TGMS automated inspection is better than a human visual inspection in detecting track defects.

Signal detection theory was used in this study to measure and compare the ability of humans and the TGMS to detect track defects. SDT has a history of use for the purpose of evaluating diagnostic systems (see Swets and Pickett, 1982; Swets, 1996). In the present case, human visual inspection was conducted by four FRA track inspectors per field test on a hi-rail vehicle using a customized push button data acquisition device. Contrary to normal practice, the TIs did not exit the hi-railer to perform a ground inspection when an anomalous track condition was spotted. A TGMS system mounted on a second hi-rail vehicle collected data from the same tracks. There were two different railroad test sites on which the inspections were conducted. The first test zone was approximately 11 miles long, and the second test zone was approximately 13 miles long. Four human exception editors (EEs) processed the TGMS data from a remote site. These EEs reviewed the exception reports from the TGMS to identify and filter out potential false alarms. In SDT it is necessary to know, independent of the observers, what is a signal (i.e., track defect) and what is noise (i.e., the absence of a defect; see Chapter 5, Table 5.1, p. 63). To determine this, an independent consultant with an extensive background in track inspection was used to conduct a comprehensive track inspection at both field test locations. The inspection took two days to complete at each location. Track defects which were detected by the TIs or TGMS were Hits, while detections of non-defects were False Alarms (FAs). Track defects which were not detected were Misses. Correct Rejections were "…calculated by subtracting the sum of Hits, FAs, and Misses from the total number of 160-foot segment lengths." in each test site (Al-Nazer et al., 2017, p. 55). From this data the probability of Hits (p(Hit)) and FAs (p(FA)) were calculated and used in the comparison.

The p(Hit) and p(FA) data were used to calculate standard SDT measures of sensitivity (d') and bias (β). These measures depend on the assumption that signal and noise distributions are normal and equal in variance, so Al-Nazer et al. (2017) tested these assumptions and found that they were not supported by the data (Al-Nazer et al., 2017, Figures 5 and 6). Consequently, non-parametric measures of sensitivity (A') and bias (B'') were calculated and used in all statistical comparisons of TI and TGMS inspections (see Chapter 5 for more detail on SDT).

Figure 10.9 shows a plot of p(Hit) as a function of p(FA) for the TI and TGMS (labeled ETG) observers in the study. This encompasses all of the inspection data in the study and provides the best overall picture of the data in this study.

Figure 10.9 is known as a Receiver Operator Characteristic (ROC) plot in SDT. Two things are suggested by this figure. First, TGMS Observers appears to have a greater bias to say "NO" regarding the presence of track defects than TI observers. Second, the TGMS observers appear to have greater sensitivity than the TI observers.

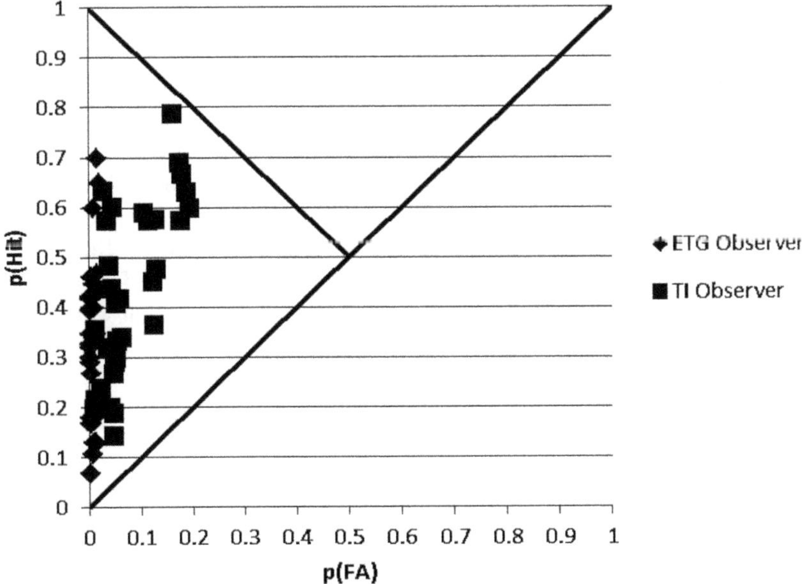

FIGURE 10.9   ROC plot for TI and TGMS (labeled ETG) observers.

Al-Nazer et al. (2017, Figure 17) show that the bias to say "NO" (B") is statistically higher for the TGMS observers but A' is not statistically different for TGMS and TI (Al-Nazer et al., 2017, Figure 11).

Al-Nazer et al. note that the TGMS observers, in particular, seem to be setting the criterion for their decision-making as a Neyman-Pearson observer. A Neyman-Pearson observer would set a fixed FA rate and maximize the Hit rate within that limitation. The mean FA probability for the ETG observers was 0.007 ± 0.002. For the TI observers the mean FA probability was 0.08 ± 0.02. Another possible factor contributing to the criterion being set to favor low FA probabilities for TGMS and TI observers is that the prior odds (probability of no defect/probability of a defect, p(noise)/p(signal)) is very high in both groups (3.41 and 3.22 for TGMS and TI, respectively). This means that defects are infrequent (probability of a defect is 0.23) and the absence of a defect is high (probability is 0.76). Bias is driven, in part, by prior odds (see Chapter 5, Equation 5.4). Bias is also determined by the values placed on the decision outcomes. It is possible that FAs carry a high negative value by policy and are avoided by track inspectors for that reason. However, a decision to keep FAs low carries the consequence of a high Miss rate. Consequently, a missed defect could result in an accident. Policy concerning FAs should take all the costs and benefits of decision outcomes into consideration to improve the effectiveness of track inspection whether by human or machine.

It also needs to be noted that TGMS is not a single system. "There are several suppliers of geometry systems throughout the industry, and they each use unique technologies. Some systems are based on integrating and filtering of inertial sensor data, while others rely on mechanical means to measure mid-chord offsets (MCOs) then convert the MCOs to track geometry parameters such as alinement and profile with various algorithms." (Al-Nazer et al., 2017, p. 5). Moreover, as noted by France-Peterson et al. (2023), the role of the human operator must also be considered.

The study by Al-Nazer et al. (2017) is limited with regard to the range of TGMS systems examined (one) and the role of the human operator within that system (data collection with human present to monitor; human review and processing of automated data from a remote location; human decides on and executes an action, such as a repair or speed restriction). France-Peterson et al. (2023) note that there are at least three options within each of these three aspects (data collection, data analysis, decision-making) of the automated and human system. Moreover, the role of the track inspector (TI) in the study is also limited in that the TIs would normally stop the hi-rail vehicle "…to conduct a more thorough ground inspection when they spot a potential anomalous track condition from within the hi-railer." (Al-Nazer et al., 2017, p. 8). These are important limitations to consider regarding the effectiveness of human vs. automated track inspection.

It is clear that the Al-Nazer et al. study is just a beginning point in comparing the effectiveness of different types of track inspection systems. What this study strongly suggests is that a controlled environment in which track defects of known characteristics are used, and in which bias is controlled and varied by the experimenters, would produce more easily interpreted test outcomes.

NOTES

1 A hi-rail vehicle is a self-propelled vehicle that is equipped with retractable flanged wheels so that the vehicle can legally be used on both roads and rails.
2 Track miles refers to the total number of miles of track. A territory of 10 miles with double track has 20 track miles.
3 Tangent track is straight track in railroad parlance.
4 Jointed rail has segments that are attached by joint plates.
5 Continuous welded rail.

REFERENCES

Al-Nazer, L., Raslear, T., Patrick, C., Gertler, J., Choros, J., Gordon, J., & Marquis B. (2011). *Track inspection time study* (Report No. DOT/FRA/ORD-11/15). Washington, DC: U. S. Department of Transportation.

Al-Nazer, L., Raslear, T., & Welander, L. R. (2014). *Rail integrity application of ultrasonic phased arrays for rail flaw sizing* (Report No. DOT/FRA/ORD-14/07). Washington, DC: U. S. Department of Transportation.

Al-Nazer, L., Raslear, T., Wilson, R., & Kidd, J. (2017). *Quantification of the sensitivity of two prevalent track inspection systems* (Report No. DOT/FRA/ORD-17/03). Washington, DC: U. S. Department of Transportation.

Bartley, S. H. (1951). The psychophysiology of vision. In S. S. Stevens (Ed.), *Handbook of experimental psychology* (pp. 921–984). New York: Wiley.

Boff, K. R., & Lincoln, J. E. (Eds.). (1988). *Engineering data compendium. Human perception and performance.* Wright-Patterson Air Force Base, Ohio: Harry G. Armstrong Aerospace Medical Research Laboratory.

Boyce, P. R. (1997). Illumination. In G. Salvendy (Ed.), *Handbook of human factors and ergonomics* (pp. 858–890). New York: Wiley.

Egan, J. P. (1975). *Signal detection theory and ROC analysis.* New York: Academic.

France-Peterson, M., Multer, J., & Melnik, G. (2023). *Human-automation teaming in track inspection* (Report No. DOT/FRA/ORD-23/07). Washington, DC: U. S. Department of Transportation.

Graham, C. H. (1951). Visual perception. In S. S. Stevens (Ed.), *Handbook of experimental psychology* (pp. 868–920). New York: Wiley.

Gertler, J., & Viale, A. (2006). *Work schedules and sleep patterns of railroad maintenance of way workers* (Report No. DOT/FRA/ORD-06/25). Washington, DC: U.S. Department of Transportation.

Gertler, J., DiFiore, A., & Raslear, T. (2013). *Fatigue status of the U.S. Railroad Industry.* (Report No. DOT/FRA/ORD-13/06). Washington, DC: U.S. Department of Transportation.

Graham, C. H. (1951). Visual perception. In S. S. Stevens (Ed.), *Handbook of experimental psychology* (pp. 868–920). New York: Wiley.

Green, D. M., & Swets, J. A. (1966). *Signal detection theory and psychophysics.* New York: Wiley.

Hursh S. R., Balkin, T. J., Miller, J. C., & Eddy, D. R. (2004). The fatigue avoidance scheduling tool: Modeling to minimize the effects of fatigue on cognitive performance. *SAE Transactions, 113,* 1, 111–119.

Luce, R. D. (1986). *Response times.* Oxford: Oxford Press.

Mackworth, N. H. (1950). *Researches on the measurement of human performance* (Report No. 268). London: Medical Research Council.

Rail Safety Improvement Act of 2008, Track inspection time study. Public Law 110–432, Div. A., Sec 403.

Riggs, L. A. (1971). Vision. In J. W. Kling & L. A. Riggs (Eds.), *Woodworth & Sclosberg's experimental psychology* (pp. 273–314). New York: Holt, Rinehart and Winston.

Swets, J. A. (1996). *Signal detection theory and ROC analysis in Psychology and diagnostics. Collected papers*. Mahwah, NJ: Erlbaum.

Swets, J. A., and Pickett, R. M. (1982). *Evaluation of diagnostic systems. Methods from signal detection theory*. New York: Academic.

Wickens, T. D. (2004). *Elementary signal detection theory*. New York: Oxford Press.

Index

Note: **Bold** page numbers refer to tables and *italic* page numbers refer to figures.

For Product Safety Concerns and Information please contact our EU
representative GPSR@taylorandfrancis.com
Taylor & Francis Verlag GmbH, Kaufingerstraße 24, 80331 München, Germany

www.ingramcontent.com/pod-product-compliance
Ingram Content Group UK Ltd.
Pitfield, Milton Keynes, MK11 3LW, UK
UKHW022315100726
473146UK00009B/485